TREATISE

ON THE

MECHANICAL THEORY OF HEAT

AND ITS APPLICATIONS

TO THE

STEAM-ENGINE, ETC.

BY

R. S. McCULLOCH,

CIVIL ENGINEER AND
PROFESSOR OF MECHANICS IN WASHINGTON AND LEE UNIVERSITY, LEXINGTON, VA.

NEW YORK:
D. VAN NOSTRAND, PUBLISHER,
23 MURRAY AND 27 WARREN STREET.
1876.

Electrotyped by SMITH & McDOUGAL, 82 Beekman Street, New York.

TO

PROFESSOR JOSEPH HENRY,

AS A

TOKEN OF ESTEEM AND AFFECTION,

THIS

BOOK IS GRATEFULLY INSCRIBED.

PREFACE.

OF part of this book, privately lithographed for my pupils, a few copies given to scientific friends caused some to express a desire for its publication. Hence, after revision, it is offered to the public, with the hope that it may prove useful and acceptable.

Any one acquainted with only the elements of analytical geometry, and of the fluxional calculus, should find no difficulty in understanding all it contains.

In this country, however, scientific education, as well as classical, has unfortunately retrograded; and superficiality is the fashion of the day. Hence, some anxious for scientific knowledge, with the least labour and in the shortest time, imagine it might be well in scientific literature to dispense with the calculus. To them no better advice can be given than to begin by studying it thoroughly, if they would reasonably hope ever to comprehend much which must otherwise be unintelligible.

In this book, as in all his instruction to young men, it has been the effort of the writer to keep steadily in view the sublime unity, simplicity, and perfection of those laws which are manifest in the obedient physical universe. Laws which could not exist without a Law-giver. To enable us thus to see more clearly the omnipotence and wisdom of God revealed in his works is certainly the true and highest office of human science. And such studies are also an intellectual preparation most fit for the docile acceptance of that Christian faith, which, we are told, is the substance of things hoped for and the evidence of things not seen.

Sources of information have generally been indicated; but it is difficult to avoid their accidental omission, when intent chiefly upon demonstration. And in historical matters especially, it is almost impossible to be perfectly accurate and just. Yet the writer is unconscious of failure in these respects.

Lastly, he gratefully acknowledges his indebtedness to the profound views of his friend Prof. W. H. C. Bartlett, whose mathematical exposition of the unity of physical action has been the point of departure of his own labours.

CONTENTS.

PART I.

CHAPTER I.

HISTORICAL.

CHAPTER II.

DYNAMICS.

CHAPTER III.

DYNAMICS.

CHAPTER IV.

GENERAL LAWS.

CHAPTER V.

AIRS AND VAPOURS.

CHAPTER VI.

INTERNAL ENERGY.

CHAPTER VII.

AIR ENGINES.

CHAPTER VIII.

THERMAL LAWS.

PART II.

APPLICATIONS.

CHAPTER IX.

INTRODUCTION.

CHAPTER X.

STEAM AND VAPOURS.

CHAPTER XI.

STEAM-ENGINES, THEIR DEFECTS AND IMPROVEMENT.

CHAPTER XII.

MISCELLANEOUS APPLICATIONS.

PART I.

MECHANICAL THEORY OF HEAT.

CHAPTER I.

HISTORICAL.

1. The mechanical theory of heat, sometimes called thermodynamics, is that branch of science which treats of the phenomena of heat as effects of motion and position.

Thus defined, it is of recent development, and is not only interesting in itself, but of great practical importance. For by it we are enabled to correctly understand the steam and other engines, to calculate their efficiency, and appreciate their imperfections.

2. The mere speculative idea, that heat and light may be motion, is found in the writings of ancient as well as of modern authors. But imaginations are of no value and of little merit, so long as they remain barren of positive results.

Real knowledge upon this subject dates only from the year A. D. 1690, when Huyghens, in his celebrated "Traité de la Lumière," published his truly elegant demonstrations of reflexion, refraction, and double refraction, regarded as phenomena of wave propagation.

Unfortunately, not only the theory, but also the well-observed facts of Huyghens were rejected by his great cotemporary, Sir Isaac Newton; the influence of whose erroneous example seems to have fettered the minds of his followers for more than a century, and thus to have most sadly obstructed the progress of knowledge. Of all the eminent men who, during the eighteenth century, labored so successfully to extend Newton's astronomical discoveries, Euler alone appears to have had the requisite independence of thought to appreciate and adopt the investigations of Huyghens; but he was too busy with his marvellous labours in mathematics to do much to extend our knowledge of light and heat.

During all that century, the false Newtonian hypothesis, that light and heat are matter attracted or repelled by other matter, with forces analogous to gravitation or to chemical affinity, swayed the minds of scientific men. And Dr. Black's important discoveries of latent heat and of the chemical decomposition of alkaline carbonates, the latter of which soon led to the great revolution in chemistry achieved by Lavoisier and his associates, contributed much to confirm that erroneous belief.

When, at the beginning of the present century, Young in England, and soon afterwards Fresnel in France, resumed the chain of discovery so happily begun by Huyghens, it was only against strenuous and sometimes even bitter opposition that they could obtain consideration for their valuable researches. Every experimental fact was scrutinized with skeptical suspicion, trifling imperfections were exaggerated into fancied contradictions, and evidence the most conclusive was often rejected without fair examination. On the other hand, the cherished hypothesis, that heat and light are matter, was overloaded with postulates the most preposterous, for the purpose of still reconciling it with the progress of experimental discovery.

The splendid memoirs of Fresnel, recently collected and published by the French government, as a fit tribute to his memory and their own intrinsic value, were by the scientific men of his day, with few exceptions, among whom Arago and Ampère should ever be remembered, cast aside, despised, and allowed to go without attention or publication—some of them even to be lost for years, until hunted up among mislaid papers.

Had Fresnel been ambitious of power, position, or praise, such unjust treatment might well have disheartened or driven him from his glorious work; but amiable, modest, and sincerely earnest, he loved truth for its own sake, and labored indefatigably in its investigation. Subsequent generations have reversed the decisions of his contemporaries, and now the scientific world points with admiration to the name of Fresnel, as that of one than whom none worthier ever earned the wreath of immortality.

This is, indeed, a dark picture for contemplation, one of human weakness, of the proneness of even the ablest of our race to persist in the blind folly of prejudice, but it is as instructive as it is sad.

So intimately connected are light and radiant heat, so precisely similar are the phenomena of both, in reflexion, refraction, polarization, and interference, that when, as in case of the sunbeam, they come together from the same source, and act in the same manner, it seems scarcely sufficient to call them analogous; and, with Melloni, we are compelled to pronounce them physically identical; differing chiefly in the distinct physiological sensations they produce in us, and therefore varying not more, perhaps even less, than violet light does from red. The triumph of the Huyghenian theory of light was, consequently, the simultaneous establishment of the mechanical hypothesis of heat.

Radiant heat can be best studied in close parallel comparison with the phenomena of optics. It is not, however, to those subjects, but to that of heat applied as power, or energy, to do work, as in the steam engine, that your attention is here invited.

3. In the year 1798, Count Rumford published experiments on the large amount of heat produced by friction in boring cannon. He observed particularly the fact that its source appeared "evidently to be inexhaustible," and logically argued that "anything which an insulated body, or system of bodies, can continue to furnish *without limitation*, cannot possibly be a material substance," and that it is "extremely difficult, if not impossible, to form a distinct idea of anything capable of being excited and communicated in the manner that heat was excited and communicated in these experiments, except it be motion."

Rumford also sought to determine the ratio of heat to the mechanical work requisite to its development by friction, or what is now called the *mechanical equivalent* of heat. He found that the work of one horse during two hours and a half is sufficient to raise through 180° Fahr. 26.58 pounds of water. From which it may be calculated that one pound heated one degree is equivalent to 940 British units of work. No allowance was made for loss of heat by radiation, and the result is, therefore, too high; this was, however, indicated as an imperfection by Rumford himself, and it amounts to about 20 per cent. These admirable experiments constitute the first positive step in thermodynamics; and for the time when they were made, as well as for the perfectly clear and logical theoretical views deduced from them, they are remarkable.

Rumford omitted to prove, by actual experiment, that the capacity for heat of metallic chips and powder produced in boring cannon does not differ perceptibly from that of unbroken metal. It was therefore contended that such might be the fact; and to explain his results in accordance with the material hypothesis, latent heat was supposed to have been given out by the disintegrated metal.

Consequently, Sir Humphrey Davy, in 1799, performed the crucial experiment of melting lumps of ice by rubbing them together, when both their own temperature and that of the surrounding bodies was below the freezing point of water. It could not possibly be assumed that ice gives out latent heat in becoming water; and if heat developed in friction be imagined to be matter, the conclusion is inevitable that in this experiment of Davy matter is *created,* which is simply incredible. Hence Davy concluded that "heat is motion, and the laws of its communication are the same as those of the communication of motion." Except by Young, these views of Davy and Rumford were long neglected in England, and generally considered to be rather speculative.

4. In the year 1807, Fourier communicated to the Institute of France the laws of the transmission of heat by radiation and conduction, subsequently published in his "Théorie Analytique de la Chaleur," and laid the foundation for the mathematical theory of heat.

Sadi Carnot, son of the famous General Carnot, published in 1824 a work entitled, "Réflexions sur la puissance motrice du feu," in which he compared the potential energy of heat to that of a *dynamic head,* or fall of water, from one level to another, and announced the very important law, now called the *theorem of Carnot,* that the greatest possible amount of work which can be performed by any heat engine is a function solely of the limits of temperature, or *chute de chaleur,* between which the engine works, and does not depend at all upon the nature of the substance heated. He showed that this substance is only passive, transmitting power as a rope does. When, therefore, the transmission takes place without waste of heat, the work will be a maximum. An engine thus supposed to work without waste, between two limiting temperatures, would be theoretically perfect, but practically such an engine is an impossibility.

Excellent as the dynamical views of Carnot are, he was yet led by the material hypothesis into the serious error of supposing the quantity of heat received from the fire equal to that delivered to the refrigerator, if used without waste during the *chute de chaleur ;*

whereas its amount requires to be diminished by all the heat transformed into mechanical work by the engine. This error pervades all writings and discussions relative to the power of heat, or steam, in which the material hypothesis is employed; for heat transformed into work would, according to that hypothesis, be an annihilation of matter, and therefore physically impossible. Although Carnot refrained from fully believing the material hypothesis and regarded it rather skeptically, he yet was misled by it, and consequently failed to determine the form of the function whose existence and importance he had discovered.

The profound views of Sadi Carnot, set forth somewhat obscurely in his book, were in 1834 put into the definite and clearer form of analytical expressions and geometrical diagrams by Clapeyron; who availed himself of the diagram of work or energy, drawn by the indicator of Watt, to show how a cycle of Carnot should be geometrically represented.

5. The idea that heat and mechanical work, or energy, are mutually and definitely convertible, appears to have occupied the minds of several persons at nearly the same time. In France, Seguin in 1839; in Germany, Mayer in 1842; in Denmark, Colding in 1843; and in England, Joule from 1843 to 1849; —each was independently led, by similar thought and reasoning, to determine and publish measurements of the equivalence of heat and mechanical work. Rumford also, as we have stated, had previously obtained for the same, in 1798, an approximate value. Reduced to the common standard of French units of work, their respective results give, for one *calorie*, the following mechanical equivalent values in work or energy:

Rumford	515	kilogrammetres.
Seguin	650	"
Mayer	365	"
Joule	425	"

Of these observers, Joule deserves to be regarded as the one to whom we are most indebted. For his laborious and faithful experiments, repeated in various ways, during a period of several years, and always with the greatest skill and care, furnish the most reliable and complete data we yet possess for the determination, not only of the mechanical equivalent of heat, but also of many other thermal effects. It is worthy of notice that the result of Rumford, when compared with that of Joule, appears to be a very close approximation, proper correction or allowance being made for radiation, conduction, etc.

Equal merit with that of Joule has been claimed for Mayer; but while he published speculations and experimented imperfectly, thereby causing ideas intrinsically valuable to be looked upon as visions, Joule labored to verify every probable conjecture by exact experiment. Consequently, his results commanded more confidence and respect, especially when the confirmation of some of the more important by Regnault had placed their accuracy beyond question. To Joule, therefore, or rather to his admirable investigations, must be justly awarded the superior merit of having caused the true theory of heat, so long disregarded and rejected, to meet finally with general reception.

6. That reception, however, required, before it could be fully accorded and thermodynamics could take its appropriate rank as a recognized part of exact science, that another and a very different labour should also be thoroughly performed, to wit, the mathematical application of the laws of energy to the exact investigation of the well-known thermal phenomena. This was a task of no slight difficulty and magnitude, but it has been admirably performed by Sir W. Thomson and Rankine, in Glasgow, and by Clausius, in Zurich, each working independently of the others.

7. Even if convenient, it would not be desirable to follow, in strict chronological order, the steps of their mathematical

discussion of our subject. Published in transactions of learned societies, or in scientific journals, during a period of several years before and after 1850, many of the results still remain scattered and in their original form.

Rankine has published several editions of his valuable treatise on the steam-engine; but unfortunately, like other books of that able and eminent engineer, it is written in a style so brief that beginners find it obscure. It has, however, the merit of having been the first systematic treatise ever published on the steam-engine in which it was explained in accordance with the true theory of heat as power or energy.

The valuable and beautiful popular lectures of Tyndall, and a small volume by Balfour Stewart, are works of real merit, well calculated to eradicate false notions, to excite interest, and to diffuse correct elementary knowledge. And the essays of Prof. Tait, which originally appeared in the *North British Review*, but have since been published in separate form, constitute an excellent historical sketch, but they neither profess, nor were intended, to meet the demands of the professional student. So also with reference to the more recent popular treatise by Prof. C. Maxwell, it does not supply what is needed by him.

Consequently, we propose here to give information which will not be found in English treatises; but will not attempt more than an elementary outline, excluding all that is hypothetical, and even much which though positive is but imperfectly developed, and referring for more extended knowledge to original memoirs and to compilations in other languages.

8. As descriptions of both the experiments and the apparatus employed by Joule for the determination of the mechanical equivalent of heat are given in most of the recent treatises on physics, we will simply refer to them for such details; but with the remark, that of his latest and most perfect data, those of 1849, the mean

result 425, obtained by the agitation of water and mercury, is the most probable value.

We consequently adopt the number 425 kilogrammetres, and call it, as is now usually done, Joule's mechanical equivalent of heat. To obtain this number from the measures used by Joule, British feet and pounds must be reduced to French metres and kilogrammes, and temperatures Fahrenheit to those of the Centigrade scale.

Denoting Joule's equivalent by the letter E, and by Q the quantity of heat measured in thermal units, called *calories*, then

$$EQ = \Sigma \int P dp$$

is the analytical expression for the first law of thermodynamics, or the law of Joule, as it is often called in honor of him to whom we chiefly owe its experimental investigation.

9. To complete the work of Joule, one step remained unfinished. He had determined the heat produced by a given amount of mechanical work; the solution of the inverse problem, that of measuring the work done by heat, was still wanting. It was accomplished by G. A. Hirn, an eminent engineer of Colmar, in Alsace. It is quite impossible to give a brief and adequate account of his admirable investigations; they must be read in their original form and language.

The steam-engines of 100 horse power in the large manufacturing establishment of Haussman, Jordan, Hirn & Co. were the apparatus used in the experiments. The water and fuel supplied to the boilers; the temperature and elastic force of the steam generated and conveyed to the cylinders; the expansion of the steam and its pressure upon the piston, represented by diagrams traced with the indicator of Watt, or automatically by the engine itself; the temperatures of the condensed steam and of the water of refrigeration; the loss by radiation and conduction: these, and all other observable data had to be measured accurately; and the difficulty of such measurements can readily be imagined. The final result was, in each case, the numerical determination of the amount of work done and of the two quantities of heat required by the law of Carnot, Q_2 that received by the steam from the fire, and Q_1 that given off in part to the condenser and in part lost by radiation and conduction. Their difference is the amount of heat corresponding to the work performed and gives

$$E(Q_2 - Q_1) = \Sigma \int_1^2 Pdp.$$

Evidently, this expression should give for E the same numerical value 425 as that found by Joule;—the actual value obtained by Hirn was 415; differing, therefore, by two per cent. But when we consider the extreme difficulty of measuring such quantities as the heat lost from a large engine, this result appears, not as a discrepancy, but as a complete confirmation. Closer approximation could not be expected or desired.

Prior to the reception of the mechanical theory of heat, it was generally held that the quantities Q_2 and Q_1 are equal; for heat, if material, should be indestructible. Such a supposition reduces the first member of the last equation to zero; and the material hypothesis consequently involves the absurdity of work done without expenditure of energy. Nothing could show more conclusively how that hypothesis must have obstructed true knowledge than this result that, according to it, the steam-engine becomes a realization of perpetual motion. Happily, the experiments of Rumford, Davy, Joule, and Hirn, have put an end forever to ideas which lead to such an absurdity.

It may render this notice of Hirn's experiments more satisfactory, if we briefly consider the cycle of operations which take place in a condensing engine, when they have become regular, or periodically constant, and indicate how they give the quantities Q_2 and Q_1.

A quantity of water w is taken from the condenser and forced by the supply pump into the boiler; where, at the higher temperature t_2, it is converted into steam, and then passes into the working cylinder. To accomplish this change from liquid water into steam, a certain quantity of heat

$$Q_2 = Lw$$

must be received from the fire. The coefficient L is called the latent heat of evaporation; it expresses the heat requisite to convert a unit of water into steam of the given temperature and elastic force, and it has been very exactly determined by Regnault. In the cylinder the steam expands, pushes the piston, is partly condensed, and then is forced by the engine back again into the condenser. There it is entirely converted into water of the original temperature, t_1, giving off in its condensation an amount of heat sufficient to raise the water of refrigeration w' from a temperature t_0 to t_1. If we denote this quantity by

$$w'(t_1 - t_0),$$

and the heat lost by radiation by h, then will

$$Q_1 = w'(t_1 - t_0) + h.$$

The work done during the cycle of operations by the engine was carefully determined by the indicator of Watt and the methods usually employed for calculating the work of machines. We have, therefore,

$$E(Q_2 - Q_1) = \Sigma \int_1^2 Pdp,$$

in which the only unknown quantity E is determined by the values experimentally found for the others.

To every intelligent mind there must be pleasure in beholding truths of nature thus beautifully investigated, and the steam-engine, that great invention of the last century, thus finally made by this eminent French engineer to bear true testimony to the law, that work done is always power expended.

We shall see that the error of assuming the heat received and given off, Q_2 and Q_1, to be equal was not the only serious one relative to the steam-engine corrected by the dynamical theory and the researches of Hirn. Well, therefore, has he earned the right to be honored as one of the few who have done most to promote true knowledge upon this important subject.

10. We cannot better conclude this brief and very incomplete historical sketch, than by showing how very imperfectly the steam-engine was understood only a few years ago.

In the celebrated treatise of De Pambour, entitled, "Théorie de la Machine à Vapeur," published in 1844, then far superior to other works, and still in many respects one of the very best standards on the subject, we find repeated and endorsed (see p. 84, *op. cit.*) the erroneous idea of Watt, that the sum of the free and latent heat of saturated steam under any pressure is a constant quantity. This error was first corrected by Regnault, in 1847, though others had previously doubted it without ascertaining the truth; in his very laborious and exact investigations made for the French government, and published in the memoirs of the Institute, Vol. XXI, 1847, he obtained for the total heat of evaporation the formula since generally adopted and used,

$$\theta = 606.5 + 0.305t.$$

"According to this formula (we translate from the original, page 727), the total heat contained in a kilogramme of saturated steam at the temperature t is equal to the quantity of heat which a kilogramme of saturated vapour at 0° gives off in becoming water at 0°, increased by the product $0.305t$. The fraction 0.305 is the quantity of heat which must be furnished to a kilogramme of saturated vapor to elevate its temperature 1°, when at the same time so compressed as to maintain its state of saturation."

Such was the extent of knowledge which caused Regnault, at the close of his labours, to criticise the steam-engine and pronounce it grossly imperfect.

Let us follow his calculations, by applying them to the example of Hirn's engines, and then compare the result with their actual work.

Suppose the boiler of an engine to generate steam of the temperature 146° and the condenser to be at 34°; then, by Regnault's formula, the total heat of evaporation is in calories

$$Q_2 = 606.5 + 0.305 \times 146 = 651.$$

And this steam condensed at 34° would give off

$$Q_1 = 606.5 + 0.305 \times 34 = 617.$$

The difference, Q_2 less Q_1, or 34 calories, would be all the heat which can be converted into work out of 651 calories received from the boiler. Hence the maximum coefficient of such an engine should be only the ratio of 34 to 651, or one-nineteenth nearly.

Now the actual working results obtained by Hirn from four of his engines were:

No. 1.	Single cylinder,	between	149°	and	31°,	efficiency,	$\frac{1}{8}$.
" 2.	"	"	149	"	25,	"	$\frac{1}{10}$.
" 3.	Woolf.	"	143	"	41,	"	$\frac{1}{7}$.
" 4.	"	"	143	"	39,	"	$\frac{1}{8}$.

The mean of which results is the efficiency of $\frac{1}{8}$ for the limits of temperature 146° and 34°.

We see that even the least advantageous of these experimental trials gave an efficiency nearly twice as great as was then held to be possible, according to the theoretical views of every eminent engineer of that day. Thus, therefore, in full contradiction of their accepted ideas, the steam-engine proved itself to be twice as powerful or perfect as was generally supposed to be possible.

Here, indeed, was a wide discrepancy between theory and actual fact, not easily to be explained away. It could not be ascribed to error of experiment, for Regnault's investigations communicated to the Institute, copied into scientific journals, and scrutinized in their most minute details, were models of skill and marvellously exact; neither could the results of Hirn be doubted.

This conflict was soon happily removed by an important discovery, made partly by Rankine and partly by Hirn. It had long been known that water accumulates in and obstructs the cylinders of engines, and this water was erroneously believed to be accidentally carried by the steam from the boiler; Rankine gave the true explanation, in 1849, by ascribing the presence of water in the cylinder to the condensation of steam working and losing heat during expansion.

Steam had previously been supposed to enter into and pass out from the cylinder entirely in the state of saturation; but for this assumption there was no proof whatever. If such be not the case, then it must pass out either partially condensed or superheated. To suppose it to be superheated is only to increase the discrepancy, by diminishing the difference of heat and corresponding work. The only remaining hypothesis is therefore the true one; it is in fact partially condensed, and passes from the cylinder as a mixture of steam and liquid water, having transformed a large part of its latent heat into mechanical work.

But, though Rankine thus gave the true explanation of this exceedingly important fact, its actual experimental demonstration was first made by Hirn some years afterwards. Having, for this purpose, attached to one of his engines, working with an elastic force of five atmospheres, a metallic pipe into which glass plates were tightly fitted, he could see the transparent steam of 152° become clouded with liquid vesicles whenever expansion was made to take place.

Hence it follows that the latent heat of steam condensed into water in the cylinder of an engine does a large part of the work. And according to the dynamical theory, this latent heat as energy is simply transformed into mechanical work; while by the material hypothesis it must be absurdly supposed to be matter annihilated.

The apparent discrepancy between the equally exact and admirable experiments of Hirn and Regnault vanishes; for a new discovery casts light upon the subject; and a more correct theory teaches, that to understand the steam-engine we must regard its work only as the transformation of potential energy, or, in other words, as expenditure of that power which is so bountifully provided for use in extensive forests, and in the vast masses of underlying coal, the remains of forests which grew in wild luxuriance and were stored up during ages, long before this Earth was ready for habitation by man.

CHAPTER II.

DYNAMICS.

FUNDAMENTAL LAWS.

11. It is necessary to the comprehension of our subject that the fundamental laws of mechanics be well understood, both in their relative connexions with each other and in their thermal applications. We shall, therefore, briefly present those which we will most often need to employ, and will thus not only give a summary of the most important for your convenient reference, but also be enabled to define accurately some terms which are of frequent use in thermodynamics.

12. The general equation of energy given by Prof. Bartlett, which embraces both the law of Newton, that action and reaction are always equal and opposite, and the principle of Dalembert, that lost forces are in equilibrium,—and which may be adopted as the fundamental law of analytical mechanics, is written thus:

$$\Sigma\, Pdp - \Sigma\, m \frac{d^2s}{dt^2}\, ds = 0. \qquad \text{(A)}$$

This most important expression includes all thermodynamic action; we shall, therefore, first demonstrate and afterwards deduce from it many of its principal consequents, thermal and mechanical. Its first term denotes the elementary work done, or power expended by all the forces, positive and negative, which act upon a system of bodies; and its second term expresses the equivalent *vis viva* or kinetic energy which those forces can produce. Evidently, the equation simply expresses the equivalence of dynamical effects.

13. In analytical geometry, we have for the square of the elementary distance between any two consecutive positions,

$$(ds)^2 = (dx)^2 + (dy)^2 + (dz)^2, \qquad (1)$$

but in motion velocity must be considered, and time becomes the independent variable. The successive positions indicated by the current co-ordinates, x, y, z, being functions of time, analytical mechanics is, therefore, a geometry of four dimensions, in which every motion may be expressed by the general equation

$$s = f(t) = a + bt + ct^2 + \text{etc.} \qquad (2)$$

As the velocity at any instant is measured by the space which would be passed over in the next succeeding unit of time if that velocity were to remain constant, we have always

$$dt : 1 :: ds : v,$$

or

$$v = \frac{ds}{dt}. \qquad (3)$$

The change of velocity which a force, if it remain constant, causes in the unit of time is called the acceleration due to that force; and denoting it by ϕ, we have

$$dt : 1 :: dv : \phi;$$

or

$$\phi = \frac{dv}{dt} = \frac{d^2s}{dt^2}. \qquad (4)$$

It is evident that these equations are true for all possible motions.

Integrated under the supposition that ϕ is constant, or the motion uniformly varied, equation (4) gives

$$s = s_0 + v_0t + \tfrac{1}{2}\phi t^2. \qquad (5)$$

The arbitrary constants s_0 and v_0 denote the initial space and velocity, when t is zero.

If the velocity v be constant, then

$$s = s_0 + vt \qquad (6)$$

is the law of uniform motion.

14. Forces are proportional to their effects, and, therefore, are measured by the accelerations they produce in the velocity of the same body. Hence, if we denote by w the weight of a body, by m its mass, and by g the acceleration due to the earth's force of gravity, we shall have

$$P : w :: \phi : g,$$

consequently,

$$P = \frac{w}{g}\phi = m\phi. \qquad (7)$$

If we combine this result with equation (4), we have

$$P = m\frac{dv}{dt} = m\frac{d^2s}{dt^2}. \qquad (8)$$

The effect exerted by a force is the statical pressure or strain at that instant, and may vary or not with lapse of time; but always its *instantaneous* value is given by the equation just found. For it is not necessary that the force shall actually produce the change of velocity, or even continue to act, but only that the effect would be that indicated if the action upon the body free to move should continue unchanged during the succeeding unit of time. It is highly important that this explanation of what is the instantaneous value of a constantly varying quantity be clearly and fully comprehended.

15. Forces are sometimes exerted during intervals of time, producing changes of velocity in bodies free to move, the accumulated effect is then the time-integral

$$\int Pdt = m\int\frac{dv}{dt}dt = mv, \qquad (9)$$

no constant of integration being added if v is nothing when t is zero. This time-integral is usually called the momentum, the quantity of motion, the impulsion, or *vis mortua.*

Generally, forces act upon bodies resisting motion, or reacting by inertia, cohesion, attraction, or repulsion; and the effects produced are measured by resisted changes of position. The forces operate in space, and are said to perform work, which is determined by the space-integral

$$\int P ds = m \int \frac{dv}{dt} ds = m \frac{v^2}{2}. \tag{10}$$

The first member of this equation is by different authors variously called: power applied, work done, potential energy, sum of the virtual moments, quantity of action, etc. The second member is generally called the half of the *vis viva,* or living force, the accumulated work, or the kinetic energy.

Differentiating equations (9) and (10), the force P has the values

$$P = m \frac{dv}{dt} = mv \frac{dv}{ds};$$

which are evidently identical, for

$$v = \frac{ds}{dt}.$$

And thus simply is it proved that the memorable controversy during the last century about forces, whether they are to be measured by *vis mortua,*

$$mv,$$

or by *vis viva,*

$$mv^2,$$

was merely a war of words; for in fact a force P is not measured by either, being only a factor of which they are both products.

16. When the elementary path ds makes an angle θ with the direction of the force P, as is always the case if there be deflecting forces or surfaces, then $P \cos \theta$ is the effective component and dp is the projection of ds on the line of the force. Multiplying, therefore, both sides of equation (8) by ds, substituting for P its effective component, and reducing by the value of dp, we get the fundamental law,

$$Pdp - m \frac{d^2s}{dt^2} ds = 0. \tag{11}$$

It is evident that we may regard Pdp either as the action of $P \cos \theta$ through the length ds, or as that of P through dp, the projection of ds on the line of the force, and which is called the virtual velocity.

In equation (11) the term Pdp denotes the elementary work done, or the power to do it; for if we define work to be resistance overcome by a force P through a length p, and indicate it by the symbol Π, then

$$\Pi = \int Pdp. \tag{12}$$

We may generalize formula (11) by supposing any finite number of masses, m, m', m'', m''', etc., to be acted upon by the forces P, P', P'', P''', etc., and that these forces are themselves resultants of any components, positive or negative. And we will thus obtain the fundamental law,

$$\Sigma\, Pdp - \Sigma\, m \frac{d^2s}{dt^2} ds = 0. \tag{A}$$

If in this expression we suppose ds zero, then there is no motion, and our equation becomes the law of balanced forces, or of equilibrium,

$$\Sigma\, Pdp = 0, \tag{13}$$

by which all statical actions may be calculated.

17. Let us assume, as a truth established by induction, that a force may always be resolved into rectangular components, represented either by its projections on the co-ordinate axes, or by sides of a parallelopipedon of which the diagonal represents the force itself, then for P, making the angles α, β, γ, with the axes of co-ordinates, the components are

$$X = P \cos \alpha; \quad Y = P \cos \beta; \quad Z = P \cos \gamma. \tag{14}$$

And the relation of these components to P, their resultant, is

$$P^2 = X^2 + Y^2 + Z^2. \tag{15}$$

This mode of considering the theorem, usually called the parallelogram of forces, to be an inductive truth appears to be the most satisfactory. For it is not easy to prove a parallelogram of statical pressures to be a direct logical consequence of one of motions, which in such cases reduce to zero.

18. If equation (1) be divided by dt^2, it becomes

$$\left(\frac{ds}{dt}\right)^2 = \left(\frac{dx}{dt}\right)^2 + \left(\frac{dy}{dt}\right)^2 + \left(\frac{dz}{dt}\right)^2, \tag{16}$$

which equation (3) shows to be the same as

$$v^2 = v_x^2 + v_y^2 + v_z^2. \tag{17}$$

This result may be geometrically represented by a parallelopipedon of velocities, whose edges, v_x, v_y, v_z, are also the projections of v on the three co-ordinate axes.

As the value of v in this equation is arbitrary, we may replace it by dv; then dividing by dt^2, and reducing by equation (4), we have

$$\phi^2 = \phi_x^2 + \phi_y^2 + \phi_z^2. \tag{18}$$

In this expression ϕ_x, ϕ_y, ϕ_z, denote the three component accelerations in the directions of the axes; and we may geometrically construct the equation, either by a parallelopipedon of accelerations, or by projecting ϕ on the co-ordinate axes.

If now we multiply by the mass m, differentiate equation (16) as a function of the time, and generalize the result by summation, we obtain

$$\Sigma\, m \frac{d^2s}{dt^2} ds = \Sigma\, m \left(\frac{d^2x}{dt^2} dx + \frac{d^2y}{dt^2} dy + \frac{d^2z}{dt^2} dz\right). \qquad (19)$$

This formula enables us to resolve into components the second term of our fundamental equation of energy.

19. Let a, β, γ be the angles made by the co-ordinate axes with the line of the force P, and λ, μ, ν be those made with the direction ds; then denoting by dx, dy, dz the projections of ds, multiply both sides of the equation

$$\cos\theta = \cos a \cos\lambda + \cos\beta \cos\mu + \cos\gamma \cos\nu$$

by ds, and reducing by the values of X, Y, Z, given in equations 14, we have, by summation,

$$\Sigma\, Pdp = \Sigma\,(Xdx + Ydy + Zdz); \qquad (20)$$

a formula for resolving into components the first term of the fundamental equation (A), and which shows that the elementary work of the forces P is always equal to the sum of the elementary work of their components.

20. If we compare equations (11, 19, 20) and equate the coefficients of like quantities, we see that

$$\begin{aligned} \Sigma\left(X - m\frac{d^2x}{dt^2}\right) &= 0, \\ \Sigma\left(Y - m\frac{d^2y}{dt^2}\right) &= 0, \qquad (21) \\ \Sigma\left(Z - m\frac{d^2z}{dt^2}\right) &= 0 \end{aligned}$$

equations which formulas (8, 12) show to be those for the resolution into components of motions of translation, as well as for the statical pressures, of any dynamical system of bodies.

Let us transform these equations by multiplying the first by y and the second by x; then the first by z and the third by x; and lastly the second by z and the third by y. Now combining these products, we have

$$\begin{aligned}
&\Sigma\,(Xy - Yx) - \Sigma\, m\left(\frac{d^2x}{dt^2}\,y - \frac{d^2y}{dt^2}\,x\right) = 0,\\
&\Sigma\,(Xz + Zx) - \Sigma\, m\left(\frac{d^2x}{dt^2}\,z - \frac{d^2z}{dt^2}\,x\right) = 0, \qquad (22)\\
&\Sigma\,(Yz - Zy) - \Sigma\, m\left(\frac{d^2y}{dt^2}\,z - \frac{d^2z}{dt^2}\,y\right) = 0.
\end{aligned}$$

These are the equations for the component moments of rotation around the co-ordinate axes.

If of these components that around the axis of x be denoted by w_x, and those around the axes of y and z by w_y and w_z, then it is readily shown that for the resultant moment of rotation w there exists the relation

$$w^2 = w_x^2 + w_y^2 + w_z^2,$$

which is usually called the theorem of the parallelopipedon of rotations or moments.

21. To integrate equation (20) it is necessary that the coefficients X, Y, Z, be functions of the co-ordinates x, y, z, and that the variables be capable of separation. If these conditions be fulfilled, then

$$\Sigma \int Pdp = f(x, y, z) + C. \qquad (23)$$

Taking this integral between the positions, or configurations, (1) and (2), it gives

$$\Sigma \int_1^2 Pdp = f_2(x, y, z) - f_1(x, y, z). \tag{24}$$

If now the system should pass by a cycle, from the position (1) back again to the same state, then

$$\Sigma \int_1^1 Pdp = f_1(x, y, z) - f_1(x, y, z) = 0. \tag{25}$$

That is to say, in such a change of the system, the work done, or the energy lost or gained, will be zero.

THEORY OF MACHINES.

22. Transposing and integrating the second term of the fundamental equation, it becomes

$$\Sigma \int Pdp = \Sigma\, m \frac{v^2}{2} + C, \tag{26}$$

an equation usually called the *theorem of vis viva*, and which is of the greatest practical importance in calculating the work done by machines of all kinds.

Taking the integral between the limits or positions (1) and (2), we have

$$\Sigma \int_1^2 Pdp = \Sigma \frac{m}{2}(v_2^2 - v_1^2). \tag{27}$$

Hence, the amount of work done, or of power expended, during the change of state or position from (1) to (2) is equivalent to the corresponding variation which takes place in the *vis viva* or kinetic energy.

As work is never measured by the whole, but always by the half of the product mv^2, we shall follow the example of Coriolis in giving to the term *vis viva* the more convenient definition of the half instead of the whole of that product.

When a machine starts from repose to do work, the velocity increases until the elementary resistance R balances the power applied; the velocity then becomes a maximum, and

$$\Sigma\,(Pdp - Rdr) = \Sigma\, m\, d\frac{v^2}{2} = 0.$$

The machine now works to the greatest advantage, for power is simply converted into work, and the velocity is either uniformly or periodically constant.

If at any time we suppress the applied or motive power, then

$$\tfrac{1}{2}\,\Sigma\, mv^2 = C - \Sigma \int Rdr\,;$$

and as the second member of this equation is composed of a constant diminished continually by an increasing quantity, it must finally be exhausted. The velocity then becomes zero, or the machine stops.

CONSERVATION OF VIS VIVA.

23. If in equation (26) the forces be assumed to be only the internal mutual attractions and repulsions of the masses composing the system, then these forces, taken in pairs, being all equal and opposite,

$$\Sigma \int Pdp = \Sigma\, m\frac{v^2}{2} + C = 0, \tag{28}$$

or the sum of the *vis viva* is constant, and the system is, therefore, either at rest or in uniform motion.

This theorem is generally known as Huyghens' principle of the conservation of *vis viva*. It is evidently only of limited applicability and dependent upon the restricted conditions that there are no external disturbing forces, and that the action of the internal forces is one of mutual compensation.

CONSERVATION OF ENERGY.

24. The first members of equations (24) and (27) being identical, their second members are equal, or

$$\tfrac{1}{2}\Sigma\, m\,(v_2^2 - v_1^2) = f_2\,(x, y, z) - f_1\,(x, y, z).$$

If now in this equation we put

$$\Pi\,(x, y, z) = c - f\,(x, y, z), \tag{29}$$

the constant c being arbitrary, it becomes

$$\tfrac{1}{2}\Sigma\, mv_1^2 + \Pi_1\,(x, y, z) = \tfrac{1}{2}\Sigma\, mv_2^2 + \Pi_2\,(x, y, z),$$

an equation which may be written thus,

$$\tfrac{1}{2}\Sigma\, mv^2 + \Pi\,(x, y, z) = c. \tag{30}$$

This important transformation of equation (26) shows that there is a function Π, which if added to the *vis viva*, or kinetic energy, will give for their sum a constant value in any position of the system of bodies. This function, called by Lagrange the function Π (Mec. Anal., section III, § 25, *et seq.*), and by Green the potential function, Gauss has named *the potential.* It denotes the action dependent upon the position or configuration x, y, z, and is called by Rankine and others the potential energy of the system, a term which is likely to be universally adopted.

The theorem expressed by equation (30) may now be thus enunciated: in a dynamical system of invariable bodies, if there be no external action, and the internal forces depend only on the relative positions, or configurations, of the masses, the total energy is constant and equal to the sum of the potential and kinetic energy. Such a system is said to be dynamically conservative, and the theorem is called the principle of the conservation of energy.

It is practically impossible thus to disconnect a system from the disturbing action of external bodies; the theorem, therefore, only shows what would happen under such circumstances, imaginary and really impossible. In fact, it is never realized, and there is always dissipation of energy. But the smaller the external forces, the less will be their disturbing influence in a given time, and the nearer will the system approach, for short durations, to a theoretically conservative condition.

POTENTIAL AND KINETIC ENERGY.

25. We will now endeavour to make clear the meaning of the potential function

$$\Pi\ (x,\ y,\ z),$$

also to define more precisely the terms potential, kinetic, and total energy, and to show what is the signification of the principle of the conservation of energy.

There is power in the recoiling spring of a watch to drive its wheels; in the descending weight of a clock to give it motion; in elevated water to work mills; in burning fuel to drive steam-engines; in gunpowder to project balls; in animals nourished by vegetable food to perform labor; in zinc acted upon by acids to propel electro-magnetic engines. These are familiar instances of *potential* energy, of what Carnot named *force vive latente,* of power stored and ready, if brought into action, to be consumed or expended in doing work.

Potential energy is, therefore, but a name for the availability of forces of nature to communicate kinetic energy or perform other work. And its principal sources are: 1°, solar action; 2°, fuel or food; 3°, chemical union of reduced substances; 4°, animal effort, based upon vegetable nutrition; 5°, electricity; 6°, gravitation; 7°, elasticity.

To avoid confusion of thought and ambiguity of language, we should not use the same word to express indiscriminately an effect and its cause, work done and the power to do it. It is well, therefore, to adopt the word *energy*, first proposed by Bernouilli and afterwards used by Young, to express power to do work, or force stored and ready for use.

When a bullet shot from a gun reaches and shatters an object, overcoming its resistance and therefore doing work, it possesses *vis viva*, or kinetic energy, power previously transferred to it by gunpowder. The swiftly-descending weight of the pile-driver has energy stored up in it during its fall by gravitation; an axe cleaving wood, a fly-wheel overcoming sudden and great resistances, as in the work of crushing a mass of iron, the wind propelling ships or mills—these are all examples of energy stored in a moving body by natural forces—of power depending upon motion and therefore called kinetic energy, instead of *vis viva*, or living force, which are words without meaning.

TRANSFORMATION OF ENERGY.

26. The various forms of energy may be converted or transformed into each other. Thus, solar radiation evaporates from the sea and disperses in the atmosphere vapour of water, which descending in streams supplies power of gravitation to work mills. Solar action also stores up in growing plants potential energy of fuel and food. This fuel enables us to reduce metals from their ores; and metals consumed in voltaic circuits furnish electro-dynamic power for telegraphs, etc.

But though the different forms of power or energy appear thus convertible into each other, so ignorant are we of the nature of their modes of action, calling these as we do by undefinable names, such as electricity, chemical affinity, vitality, etc., that in the present state of science we cannot obtain the laws of convertibility for many of the forms of energy. Evidently, it is only when they may be reduced to a common measure, such as their equivalent kinetic energy or work, that they become capable of being expressed and discussed in mathematical equations.

27. The science of *energetics*, as some have proposed to name it, or theory of energy, as others prefer to call it, has not yet reached the stage of full and satisfactory development. And it necessarily follows, that speculations under titles such as correlation of forces, etc., may often be only hypotheses, useful, if at all, only to suggest inquiry.

LIGHT AND HEAT ARE ENERGY.

28. Fortunately, mathematical demonstration based upon the only solid foundation, that of many phenomena accurately observed and compared, has proved light and heat to be kinetic energy or *vis viva;* and we may now regard celestial and terrestrial mechanics, physical optics and thermodynamics each as a well-established part of that exact knowledge of force and motion which has attained to a positive progress far exceeding in depth, extent, and certainty, that of any other branch of physical science.

THE POTENTIAL FUNCTION.

29. We may now interpret equation (30), and determine the potential function. To obtain that equation, the system must be assumed to be dynamically conservative, that is to say, Σm in integration is constant, or the masses are not subject to change; the forces also do not become feeble or strong with time, but vary only with the relative positions x, y, z, of the masses. Equation (30) is therefore limited, and applicable only to such conservative systems.

To render our conceptions definite and clear, let us consider an example. Suppose a simple pendulum, attracted by the earth, to move in a vacuum without resistance, and that it oscillates to and fro in a vertical circular arc *ADB*. If now it be at its highest point of disturbance *A*, it will have a certain amount of power or potential energy of gravitation due to the state of the dynamical system for the position or configuration *A*. Falling from *A* to *D*, the lowest point of the circular arc of vibration, the potential energy becomes gradually less and at *D* is a minimum; its loss having been transformed into kinetic energy, which at *D* is a maximum. From *D* to *B* the pendulum ascends, losing kinetic but recovering potential energy. Then as it returns from *B* to *A* the phenomena recur in precisely reverse order. At the limits *A* and *B* the potential is a maximum, and the kinetic energy is zero, a minimum; but at *D*, the lowest point, or position of stable equilibrium, the kinetic energy is a maximum, and the potential is a minimum.

At all points of the path *AB* the sum of the potential and kinetic energy is, by equation (30), a constant quantity *c*, determined by the fact that at the points *A* and *B* the kinetic energy is zero and the constant *c* equal to the total initial energy, or to the maximum value of the potential Π, due to the position *A*. Calling this position or configuration of the system (1) and that for *D* (2), and denoting by the letters Π and V the two terms of the first member of equation (30), we have

$$\Pi + V = \Pi\,(x, y, z) + \Sigma\, m \frac{v^2}{2} = c,$$

and

$$\Pi + V = \Pi_1 + V_1 = \Pi_2 + V_2 = c.$$

But for the configuration *A* or (1), the value of V_1 is zero, and Π_1 is consequently a maximum; hence

$$\Pi_1 = c,$$

and

$$\Pi_2 - \Pi = V - V_2;$$

which shows that variations of potential are equal but opposite to those of kinetic energy, gain in one being loss in the other.

If the pendulum be vibrated in a resisting medium, then Σm is no longer constant, the system ceases to be conservative, and initial energy will be gradually lost in motion given to particles of the medium. Yet though energy be dissipated, it is never annihilated, but only communicated to external bodies.

30. Combining equations (23) and (29) and replacing their arbitrary constants of integration by a single constant, we find

$$\Pi\,(x, y, z) + \Sigma \int P dp = c; \tag{31}$$

which shows that, in any limited conservative system, the sum of the potential and of the work already performed is constant for all configurations of the masses, and equal to the initial or total energy, for which the function Π, or potential, is a maximum and the work done zero.

If a disturbed system seeks to return by the action of its internal forces to a state of repose or equilibrium, then at that final position the work done will be a maximum, and the potential a minimum. Hence the change in the potential may be measured by the work required to be done in passing from a disturbed state to one of equilibrium.

Taking the definite integral of equation (31) between the configurations (1) and (2), we get

$$\Pi_1 - \Pi_2 = \Sigma \int_1^2 P dp, \tag{32}$$

which shows that *the work done in passing from one state or configuration to another is equal to the variation of the potential for those states, and independent of the path followed in the change.*

It is, consequently, evident that this theorem of potential energy involves the impossibility of perpetual motion. For if in a conservative dynamical system it were possible to pass by one path or set of points from the state (1) to another (2) with less work or resistance than by another path or set of points, then by always going by one of these paths, and returning by the other, the forces would be able to produce a continual increase of energy without corresponding loss or work.

The independence of the potential of intermediate positions or paths followed, in the passage from one configuration to another, is one of its most valuable and important properties, one which renders it of the greatest utility in investigations of heat, of gravitation, of electrical and magnetic attractions or repulsions, and of other analogous phenomena.

DISSIPATION OF ENERGY.

31. We have indicated the practically impossible conditions necessary to render a system of bodies dynamically conservative (see sections 23, 24, 29). Power expended in work is generally dissipated, and recoverable only in particular cases, as when muscular effort is converted into the potential of elasticity by bending a spring, or of gravitation by lifting a weight. Sawing wood, ploughing ground, grinding corn, hammering iron, are examples of energy consumed or dissipated. Descending rivers convert the energy of their falling waters into heat by friction. A steamer quitting port for a voyage carries in her coal a definite amount of potential energy. As it burns away, the work done will be always equal to the energy of the coal consumed either usefully or wastefully. The sea cannot restore the work expended upon its resisting waves, nor can the winds give back the heated gases of the burnt coal. In the economy of nature, their carbon and hydrogen may, by solar energy, be made part of some future plant, and again become fuel or coal. But to that steamer their original energy, once expended, is dissipated or lost forever.

CHAPTER III.

DYNAMICS.

PERPETUAL MOTION IMPOSSIBLE.

32. We have obtained the fundamental laws of dynamics, and now propose to deduce some of their more important consequences, such as the impossibility of perpetual motion.

Resuming the discussion of the fundamental equation of energy,

$$\tfrac{1}{2} \Sigma\, m\, (v_2^2 - v_1^2) = f_2\,(x, y, z) - f_1\,(x, y, z), \qquad (33)$$

obtained by integration between the limits (1) and (2), it appears that, if a conservative system pass by any path or cycle from the state (1) back to the same primitive state or configuration, the two terms in the second member of this equation become identical and its value is zero. It is, therefore, impossible that any permanent change of kinetic energy, or velocity, can have taken place in the system.

But we have already proved, equation (25), that under precisely the same conditions and circumstances the work done during the cycle must be zero. It is, therefore, impossible that a limited system of masses, such as any machine set in motion and then abandoned to itself and to gravity, or to other analogous forces, such as magnetic or electrical attraction and repulsion, can do work without loss of kinetic energy and consequently of velocity. Such a moving system must therefore ultimately come to rest.

From formulas (27) and (32) combined, taking the integral for any cycle between (1) and (1), we have

$$\Sigma \int_1^1 P dp = \tfrac{1}{2} \Sigma m (v_1^2 - v_1^2) = \Pi_1 - \Pi_1 = 0,$$

which may be read thus, *work cannot be done without an equivalent expenditure of energy, either kinetic or potential;* and this is the algebraic expression of the impossibility of perpetual motion.

MOLECULAR FORCES.

33. If instead of deducing the impossibility of perpetual motion from the fundamental equation, we assume it to be an inductive truth, founded upon the proportionality of cause and effect, or admit as an axiom that an infinite amount of work cannot be done by the expenditure of a finite quantity of power, then equation (33) results as a consequence, and it may be shown (according to Helmholtz), if matter be supposed to be composed of ultimate particles, or material points, destitute of size or form, that the mutual attractions and repulsions of a system would take place in the directions of the lines between the centres of the masses and be functions of their relative distances.

As equation (33) requires that mv^2 and consequently that v^2 shall always have the same value when m occupies precisely the same position relatively to the system, it follows that v^2 is a function of x, y, z, the co-ordinates of that position, and

$$d(v^2) = \frac{d(v^2)}{dx} dx + \frac{d(v^2)}{dy} dy + \frac{d(v^2)}{dz} dz.$$

Differentiating equation (16), we get

$$d(v^2) = 2\left(\frac{d^2x}{dt^2} dx + \frac{d^2y}{dt^2} dy + \frac{d^2z}{dt^2} dz\right),$$

which equations (21) reduce to

$$d(v^2) = \frac{2}{m}(Xdx + Ydy + Zdz).$$

But the values of dx, dy, dz, are indeterminate; therefore the first and last of these equations give

$$\frac{2}{m}X = \frac{d(v^2)}{dx}; \qquad \frac{2}{m}Y = \frac{d(v^2)}{dy}; \qquad \frac{2}{m}Z = \frac{d(v^2)}{dz}.$$

Hence, if v^2 is a function of x, y, z, so also must the components X, Y, Z, be functions of the same variables or co-ordinates of position.

Suppose now the system condensed into a hypothetical material point a, the point of application of the resultant, then the action of m upon a will depend on their relative positions. But as these positions are determined by the intervening distance (r), see equation (1), or by the line joining m and a, their mutual actions will depend both in direction and intensity upon this line only. For the point a being taken as the origin of co-ordinates, we have

$$d(v^2) = \frac{2}{m}(Xdx + Ydy + Zdz) = 0,$$

whenever v^2 is a minimum, or the potential of r^2 is a maximum, so that

$$rdr = xdx + ydy + zdz = 0,$$

and

$$dz = -\frac{xdx + ydy}{z}$$

Therefore, by substitution,

$$(Xz - Zx)\,dx + (Yz - Zy)\,dy = 0,$$

independently of the values of dx and dy. Hence

$$Xz - Zx = 0; \qquad Yz - Zy = 0;$$

or the action of the material point m on a passes through a the origin of co-ordinates.

In conservative dynamical systems composed of material points, the mutual internal attractions and repulsions would therefore act in the directions of the lines joining them and vary only with their relative distances and masses.

It is, however, clear that a system, thus supposed to be composed of material points without size or form, but possessing mass, is purely a mathematical fiction. For molecules must be regarded as masses or groups of atoms or smaller particles, variously united according to unknown laws of configuration, crystalline structure, or chemical constitution; and their motions, absolute and relative, are not only translations of their centres of gravity, but also oscillations and rotations around those centres. Nor does this difficulty vanish if we seek to apply the reasoning of Helmholtz to atoms which may be supposed to compose the molecules, for even they cannot be assumed to be mathematical points destitute of size or form, acting centrally so as to produce only translation without rotation.

INTEGRABILITY OF THE FUNDAMENTAL EQUATION.

34. We have asserted, § 21, that the expression for work or energy,

$$\Sigma\, Pdp = \Sigma\,(Xdy + Ydy + Zdz),$$

cannot be integrated unless X, Y, Z, are functions of x, y, z, the co-ordinates of m; and we will now show that this equation is integrable for systems in which the mutual actions X, Y, Z, are functions of the masses and their relative distances.

Let x, y, z, and x', y', z', be the co-ordinates of any two molecules or masses m and m', and r be the distance of their centres. Also let $\phi\,(r)$ be the function of the distance which denotes their mutual action. Then the components of the action of m upon m' will be

$$\phi(r)\frac{x-x'}{r}, \qquad \phi(r)\frac{y-y'}{r}, \qquad \phi(r)\frac{z-z'}{r};$$

and those of the reaction of m' upon m are

$$-\phi(r)\frac{x-x'}{r}, \qquad -\phi(r)\frac{y-y'}{r}, \qquad -\phi(r)\frac{z-z'}{r}.$$

We have, therefore, for the work of m and m',

$$\frac{\phi(r)}{r}\left[(x-x')(dx-dx')+(y-y')(dy-dy')+(z-z')(dz-dz')\right].$$

But

$$r^2 = (x-x')^2 + (y-y')^2 + (z-z')^2,$$

and

$$rdr = (x-x')(dx-dx') + (y-y')(dy-dy') + (z-z')(dz-dz').$$

Hence, by substitution in the equation just found, and extension of the result to all the masses taken in pairs,

$$\Sigma\,\phi(r)\,dr = \Sigma\,(Xdx + Ydy + Zdz),$$

which is evidently integrable when the function $\phi(r)$ is known.

MOTION OF THE CENTRE OF GRAVITY.

35. The motion of any body, or system of bodies, may be decomposed into two motions; one common to all its molecules or masses, their translation in space referred to a fixed system of co-ordinate axes; the other, their motions relatively to each other, or to parallel but moveable axes through the centre of gravity of the system. Let us denote by x, y, z, the co-ordinates of m, one of the masses; by x', y', z', those of the common centre of gravity; and by ξ, η, ζ, the co-ordinates of m referred to the centre of gravity as a moveable origin. If the two systems of co-ordinates be taken parallel,

$$x = x' + \xi, \qquad y = y' + \eta, \qquad z = z' + \zeta.$$

Hence, by substituting these values in equations (21), reducing by the property of the centre of gravity,

$$\Sigma\, m\xi = 0, \qquad \Sigma\, m\eta = 0, \qquad \Sigma\, m\zeta = 0,$$

and observing that the masses have a common factor, and may therefore be added into one mass M, we get

$$\begin{aligned} \Sigma\, X &= \frac{d^2x'}{dt^2}\,\Sigma\, m = M\frac{d^2x'}{dt^2}, \\ \Sigma\, Y &= \frac{d^2y'}{dt^2}\,\Sigma\, m = M\frac{d^2y'}{dt^2}, \\ \Sigma\, Z &= \frac{d^2z'}{dt^2}\,\Sigma\, m = M\frac{d^2z'}{dt^2}; \end{aligned} \tag{34}$$

also, if we multiply these equations by the co-ordinates of the centre of gravity, x', y', z', as lever arms and combine the results,

$$\begin{aligned} \Sigma\,(Xy' - Yx') &= M\left(\frac{d^2x'}{dt^2}\,y' - \frac{d^2y'}{dt^2}\,x'\right), \\ \Sigma\,(Xz' - Zx') &= M\left(\frac{d^2x'}{dt^2}\,z' - \frac{d^2z'}{dt^2}\,x'\right), \\ \Sigma\,(Yz' - Zy') &= M\left(\frac{d^2y'}{dt^2}\,z' - \frac{d^2z'}{dt^2}\,y'\right); \end{aligned} \tag{35}$$

which evidently express the moments of rotation of the system about the fixed origin and axes.

The six equations just found show that the motion of a system of bodies, relatively to fixed axes in space, is the same as if all its masses were collected at the centre of gravity of the system, and that it is entirely independent of their mutual actions. This principle, usually called the conservation of the motion of the centre of gravity, is perfectly general, or applicable in all cases, no matter what may be the internal forces or disturbances.

Hence the centre of gravity of all the scattered fragments of an exploded shell continues to pursue the original path of the projectile. And the common centre of gravity of the system of the earth and moon revolves in its orbit around the sun, undisturbed by the daily rotations of the earth, by tides caused by the moon, by earthquakes, or by volcanic eruptions.

36. To find the expressions for the motion of the system relatively to the centre of gravity and the moveable axes, substitute for x, y, z, in equations (22) their values and

$$\Sigma\left(Y - m\frac{d^2y}{dt^2}\right)x' - \Sigma\left(X - m\frac{d^2x}{dt^2}\right)y'$$

$$+ \Sigma\,(Y\xi - X\eta) - \Sigma\, m\left(\frac{d^2y}{dt^2}\xi - \frac{d^2x}{dt^2}\eta\right) = 0.$$

The first two terms of this equation vanish, for the factors within brackets are zero.

Substituting now in the remaining terms the values of d^2x and d^2y, we obtain

$$(\Sigma\, m\eta)\frac{d^2x'}{dt^2} - (\Sigma\, m\xi)\frac{d^2y'}{dt^2} + \Sigma\,(Y\xi - X\eta) - \Sigma m\left(\frac{d^2\eta}{dt^2}\xi - \frac{d^2\xi}{dt^2}\eta\right) = 0.$$

But $(\Sigma\, m\eta)$ and $(\Sigma\, m\xi)$ are zero, the first two terms therefore disappear; and operating upon the other equations (22) in the same manner, we have

$$\Sigma\,(Y\xi - X\eta) - \Sigma\, m\left(\frac{d^2\eta}{dt^2}\xi - \frac{d^2\xi}{dt^2}\eta\right) = 0,$$

$$\Sigma\,(Z\xi - X\zeta) - \Sigma\, m\left(\frac{d^2\zeta}{dt^2}\xi - \frac{d^2\xi}{dt^2}\zeta\right) = 0, \qquad (36)$$

$$\Sigma\,(Z\eta - Y\zeta) - \Sigma\, m\left(\frac{d^2\zeta}{dt^2}\eta - \frac{d^2\eta}{dt^2}\zeta\right) = 0.$$

The co-ordinates of the moveable origin, x', y', z', having entirely disappeared from these equations, we see that the rotations around the centre of gravity must be independent of its position in space; so that the motion of the system about its centre of gravity is the same whether that centre be in motion or at rest.

CONSERVATION OF AREAS.

37. If the external forces which communicate motion to a system of bodies cease to act upon it, abandoning it thus to the equal and opposite actions of its masses upon each other; or if the forces X, Y, Z, act centrally, passing through the origin of co-ordinates; then in each of the equations (22) the first and consequently the second term will be zero.

Considering the first of those equations, putting the second term equal to zero, and integrating it as a function of t, we obtain

$$\Sigma\, m\,(ydx - xdy) = cdt,$$

and integrating again,

$$\Sigma\, m \int (ydx - xdy) = ct + c_1. \qquad (37)$$

The geometrical construction of the first member of this equation is evidently twice the sum of the areas swept over by the radii vectores of the masses, for

$$\int (ydx - xdy) = 2\int ydx - xy,$$

in which $\int ydx$ is the familiar formula for quadratures, and xy is the rectangle made by the co-ordinates of m.

If when t be zero, the bodies start from rest, then c_1, the constant of integration, vanishes; and the two remaining equations (22) similarly treated give like results, so that

$$\begin{aligned} \Sigma\, m\,(ydx - xdy) &= cdt, \\ \Sigma\, m\,(zdx - xdz) &= c'dt, \\ \Sigma\, m\,(zdy - ydz) &= c''dt; \end{aligned} \tag{38}$$

which show that the areas described around the component axes are proportional to the time of their description.

This is the well-known principle called the conservation of areas. Applied to planetary motions, it is Kepler's law of equal areas in equal times; and it further proves Kepler's law to be embraced in a far more general law involving the perturbations which the mutual attractions of the bodies of the solar system produce upon each other.

CONSERVATION OF MOMENTS OF ROTATION.

38. The first of equations (38) may be put under the form

$$\Sigma\, m \left(y\frac{dx}{dt} - x\frac{dy}{dt}\right) = c.$$

But x and y are the lever arms, and their first derivatives are the component velocities of the rotation about the axis of z; this equation, therefore, expresses the fact that the sum of the moments of rotation around the axis of z is constant. The other two equations of the group (38) give like expressions. Hence, the principle of areas is also called the law of the conservation of moments of rotation.

LOSS OF VIS VIVA IN COLLISIONS.

39. The principle of areas and the law of the motion of the centre of gravity show the motions caused by external forces to be independent of internal actions. Unlike the abstract and limited principle of conservation of vis viva, they are general, while it is applicable only in particular cases which in fact never really occur.

It is readily shown that, even when the forces acting upon a system are internal only, there is loss of vis viva, whenever shocks or collisions take place.

The simplest case is that of two equal masses destitute of all elasticity, attracting each other with equal forces and consequently moving with equal and opposite velocities; if they should come into collision, their motions would neutralize each other and Σmv^2, which was $2mv^2$ before collision, would become zero afterwards; there would, therefore, be total loss of vis viva.

If the masses m and m' as well as their velocities v and v' be unequal, then the motion of their centre of gravity will not be changed by collision. Let us denote its abscissa by x_1, and by x and x' those of the masses m and m', at any instant t, also assume, for greater simplicity, the motion to be in the direction of the axis of x. We have

$$(m + m')\, x_1 = mx + m'x',$$

and differentiating we find for the velocity of the centre of gravity,

$$(m + m')\frac{dx_1}{dt} = m\frac{dx}{dt} + m'\frac{dx'}{dt}$$

This velocity will not be changed by collision; denoting it by u, we have

$$(m + m')\,u = mv + m'v',$$

and

$$mv + m'v' - mu - m'u = 0;$$

which is true always and for all bodies, whether they be elastic or not.

Suppose now the masses to be destitute of elasticity, then will the difference of $\Sigma\, mv^2$ before and after collision be

$$mv^2 + m'v'^2 - mu^2 - m'u^2;$$

and if from this we subtract

$$2u\,(mv + m'v' - mu - m'u) = 0,$$

we get for the loss by collision

$$m\,(v - u)^2 + m'\,(u - v')^2,$$

which must always be a positive quantity, for it is the sum of the squares of the velocities lost and gained by the several masses.

If we suppose the masses perfectly elastic, and that the molecular forces restore entirely during expansion the work expended during compression, without dissipating any part of the potential of distortion in the form of vibrations, such as those of heat and sound, then would m suffer during compression a loss of velocity $(v - u)$ and an equal loss during expansion; its velocity after collision would, therefore, be

$$v - 2\,(v - u) = 2u - v,$$

while that of m' would gain $(u - v')$ during compression and a like amount during expansion, and would be

$$v' + 2\,(u - v') = 2u - v';$$

for the difference of $\Sigma\, mv^2$ before and after collision we have, therefore,

$$mv^2 + m'v'^2 - m(2u - v)^2 - m'(2u - v')^2,$$

which reduces to

$$4u(mu + m'u - mv - m'v') = 0;$$

consequently the sum of the vis viva is constant.

But to obtain this result, it has been necessary to suppose that no part of the vis viva is dissipated in the form of vibrations, an impossible condition never realized. For part of the work during compression is not restored and there is consequent loss of sensible motion, transformed into vibrations. In such actions as the ringing of bells it is quite evident that a large part of the energy must be expended in producing vibrations of sound. Moreover, whenever there is loss of sensible motion, even in such instances as the collision of very inelastic bodies, energy is not destroyed but transformed, partly into potential distortion and partly into vibrations of sound, heat, etc. Hence we see that sensible energy tends constantly to dissipation in the final form of imperceptible vibrations.

It is important to avoid confusion of thought, which sometimes occurs when the principle of conservation of vis viva is mistaken for the theorem of vis viva; the latter is given in equation (26), one of the algebraic forms of the fundamental law of energy, and is true for all dynamical actions; while the former, expressed by equation (28), is of very restricted applicability and never physically possible.

THEOREM OF VIS VIVA.

40. Let us resume the consideration of equation (26); comparing it with equation (A) and (19), and integrating both sides of equation (19) we obtain

$$\tfrac{1}{2}\Sigma\, m\left(\frac{ds}{dt}\right)^2 = \tfrac{1}{2}\Sigma\, m\left[\left(\frac{dx}{dt}\right)^2 + \left(\frac{dy}{dt}\right)^2 + \left(\frac{dz}{dt}\right)^2\right], \qquad (39)$$

which may be put under the form

$$\tfrac{1}{2}\Sigma\, mv^2 = \tfrac{1}{2}\,\Sigma\, m\,(v_x^2 + v_y^2 + v_z^2).$$

Hence, the vis viva or kinetic energy of any moving system of bodies is equal to the sum of its components in the directions of the co-ordinate axes.

41. It does not follow, and it is not true that, as is sometimes ignorantly asserted, if the motion of a system be decomposed into other motions in any manner whatever, the total vis viva will always be equal to the sum of the vires vivæ of the several component motions.

To render this evident let us divide the absolute motions of a system into those of translation referred to the fixed axes in space, and the relative motions of its masses referred to any parallel and moveable system of axes.

Denoting the absolute co-ordinates of the moveable origin by x', y', z' and those of the mass m by x, y, z, also the relative co-ordinates of m for the moveable axes by ξ, η, ζ we have

$$x = x' + \xi, \qquad y = y' + \eta, \qquad z = z' + \zeta;$$

and if these values be substituted in equation (39) it becomes

$$\begin{aligned}\Sigma\, mv^2 = {} & \frac{\Sigma\, m}{dt^2}\,[(dx')^2 + (dy')^2 + (dz')^2] \\ & + \frac{\Sigma\, m}{dt^2}\,[(d\xi)^2 + (d\eta)^2 + (d\zeta)^2] \qquad (40) \\ & + 2\,\Sigma\, m\left(\frac{dx'}{dt}\cdot\frac{d\xi}{dt} + \frac{dy'}{dt}\cdot\frac{d\eta}{dt} + \frac{dz'}{dt}\cdot\frac{d\zeta}{dt}\right),\end{aligned}$$

which may be put under the form

$$\Sigma\, mv^2 = \Sigma\, m\,(u^2 + \omega^2) + 2\,\Sigma\, m\left(\frac{dx'}{dt}\cdot\frac{d\xi}{dt} + \frac{dy'}{dt}\cdot\frac{d\eta}{dt} + \frac{dz'}{dt}\cdot\frac{d\zeta}{dt}\right), \qquad (41)$$

in which u is the velocity of translation of the moveable origin, and ω the velocity of any mass m about that origin; the equation evidently proves that generally the total *vis viva* is not equal to the sum of the *vires vivæ* of its component motions, but exceeds that sum by

$$\Sigma\, m \left(\frac{dx'}{dt}\cdot\frac{d\xi}{dt} + \frac{dy'}{dt}\cdot\frac{d\eta}{dt} + \frac{dz'}{dt}\cdot\frac{d\zeta}{dt}\right),$$

or by the sum of the masses into the products of the parallel component velocities for the two systems of parallel co-ordinates.

42. But if the moveable origin be taken at the centre of gravity of the system, then because

$$\Sigma\, m\xi = 0, \qquad \Sigma\, m\eta = 0, \qquad \Sigma\, m\zeta = 0,$$

the last term in equation (41) reduces to zero, and it takes the simple form

$$\Sigma\, mv^2 = \Sigma\, m\,(u^2 + \omega^2), \tag{42}$$

an important relation between the kinetic energy of translation and that around the centre of gravity in any moving system of bodies; which may be thus enunciated, the total kinetic energy of any system is equal to the sum of its energy of translation and its internal energy of motion relatively to the centre of gravity.

VIS VIVA OF VIBRATIONS.

43. The motions of a system may be divided into sensible motions easily observed, and molecular vibrations, which generally are too small to be seen and are very rapid. These vibrations produce the phenomena of sound, light, and heat, and are to us, therefore, of especial importance.

Let us denote by u the very small displacement of a particle m of any elastic body from its position of relative equilibrium, and assume the general law of elasticity given by experiment,

$$\frac{d^2u}{dt^2} = - n^2u,$$

to be true for the particular substance; in which expression n^2 is the value of the intensity of molecular elasticity, or the force of restitution for a displacement u equal to the linear unit.

Multiplying by $2du$ and integrating, we find

$$\left(\frac{du}{dt}\right)^2 = v^2 = c - n^2u^2;$$

to determine c, let a denote the maximum displacement when v becomes zero, this gives for c the value n^2a^2, and

$$\left(\frac{du}{dt}\right)^2 = n^2 (a^2 - u^2).$$

Transposing, extracting the square root, and integrating, we have

$$u = a \sin (nt + c), \tag{43}$$

the expression for a simple displacement in vibratory motion.

The sine of the arc nt goes through all its periodic values during an increment of 360° or 2π, or while t increases by $\frac{2\pi}{n}$, that is to say, during the time of a full vibration. Denoting this time by τ, we have

$$n = \frac{2\pi}{\tau},$$

and the last equation may be written,

$$u = a \sin \left(\frac{2\pi}{\tau} t + c\right). \tag{44}$$

Let us now suppose n to be constant and the displacement u resolved into component displacements ξ, η, ζ, in the directions of the three co-ordinate axes, then

$$\xi = a' \sin (nt + c'),$$
$$\eta = a'' \sin (nt + c''),$$
$$\zeta = a''' \sin (nt + c''').$$

The differentials of these components, substituted in the second member of the equation

$$\left(\frac{du}{dt}\right)^2 = \left(\frac{d\xi}{dt}\right)^2 + \left(\frac{d\eta}{dt}\right)^2 + \left(\frac{d\zeta}{dt}\right)^2,$$

give for the square of the component velocity,

$$\left(\frac{d\xi}{dt}\right)^2 = a'^2 n^2 \cos^2 (nt + c')$$
$$= 2\frac{\pi^2}{\tau^2} a'^2 \left[1 + \cos 2\left(\frac{2\pi}{\tau} t + c'\right)\right].$$

The mean value for the last term of this expression, being a periodical sum of the cosines of a continually increasing arc, must be zero. Hence, for the mean value,

$$\left(\frac{d\xi}{dt}\right)^2 = 2\frac{\pi^2}{\tau^2} a'^2.$$

Similar values for the component velocities of η and ζ, and for the resultant velocity of m, give for the *vis viva* of m, the mean value

$$\frac{1}{\tau}\int_0^\tau \frac{mv^2}{2}\,dt = m\frac{\pi^2}{\tau^2}(a'^2 + a''^2 + a'''^2),$$

or the *vis viva* of vibration is equal to the sum of the *vires vivæ* of the component vibrations.

Suppose now the body to be in motion in space, denote by x, y, z, the current co-ordinates of the normal position of equilibrium of m at any instant t, and by ξ, η, ζ, the components of its vibration about that normal position; and the variations

$$dx + d\xi, \qquad dy + d\eta, \qquad dz + d\zeta,$$

which occur during the time dt will give for the *vis viva* of m during the period θ, a short time, but comprising many vibrations, the mean value

$$\begin{aligned}\frac{1}{\theta}\int_0^\theta m\frac{v^2}{2}\,dt = \frac{m}{2}\left(\frac{dx^2}{dt^2} + \frac{dy^2}{dt^2} + \frac{dz^2}{dt^2}\right) \\ + \frac{1}{\theta}\int_0^\theta \frac{m}{2}\left(\frac{d\xi^2}{dt^2} + \frac{d\eta^2}{dt^2} + \frac{d\zeta^2}{dt^2}\right) dt \\ + \frac{1}{\theta}\int_0^\theta m\left(\frac{dx}{dt}\cdot\frac{d\xi}{dt} + \frac{dy}{dt}\cdot\frac{d\eta}{dt} + \frac{dz}{dt}\cdot\frac{d\zeta}{dt}\right) dt.\end{aligned}$$

But as the movements dx, dy, dz, are arbitrary and independent of $d\xi$, $d\eta$, $d\zeta$, we have

$$\begin{aligned}\frac{1}{\theta}\int_0^\theta m\left(\frac{dx}{dt}\cdot\frac{d\xi}{dt}\right) dt &= \frac{1}{\theta}m\frac{dx}{dt}\int_0^\theta \frac{d\xi}{dt}\,dt \\ &= \frac{1}{\theta}m\frac{dx}{dt}\left(\xi_0^\theta\right) = 0.\end{aligned}$$

The last term therefore disappears from the expression obtained, and we conclude that the total *vis viva* is equal to the sum of that due to the sensible motion, plus that of the insensible motion of vibration. It is clear that this result proved for any molecule m is true for all the molecules.

From the preceding demonstration, it follows that whenever vibrations continue during a length of time sufficient to include many of their periods and thus give

$$\Sigma\, m\xi = 0, \qquad \Sigma\, m\eta = 0, \qquad \Sigma\, m\zeta = 0,$$

the mean or total *vis viva* is capable of separation and is equal to the sum of *vis viva* of vibration plus that due to the other motions of the system.

VIBRATIONS CHANGE THE POTENTIAL.

44. To determine the effect of vibrations upon the potential, we have for it the two successive values

$$\Pi = \Pi\,(x,\ y,\ z),$$
$$\Pi' = \Pi\,(x + \xi,\ \ y + \eta,\ \ z + \zeta).$$

But as ξ, η, ζ, are very small comparatively to x, y, z, we may develope Π' by Taylor's theorem, which gives

$$\Pi' = \Pi + \left(\frac{d\Pi}{dx}\,\xi + \frac{d\Pi}{dy}\,\eta + \frac{d\Pi}{dz}\,\zeta\right)$$
$$+ \frac{1}{2}\left(\frac{d^2\Pi}{dx^2}\,\xi^2 + \frac{d^2\Pi}{dy^2}\,\eta^2 + \frac{d^2\Pi}{dz^2}\,\zeta^2 + \text{etc.}\right).$$

The mean value of the first differential or second term of this series is zero. But the mean value of ξ^2 will be

$$\xi^2 = a'^2 \sin^2 (nt + c),$$

which by virtue of the relation

$$2 \sin^2 w = 1 - \cos 2w,$$

reduces to the mean value,

$$\xi^2 = \frac{a'^2}{2}.$$

Therefore

$$\Pi' = \Pi + \frac{1}{4}\left(\frac{d^2\Pi}{dx^2}\,a'^2 + \frac{d^2\Pi}{dy^2}\,a''^2 + \frac{d^2\Pi}{dz^2}\,a'''^2 + \text{etc.}\right).$$

Consequently the mean value of the potential of m is changed by vibrations from that due to its normal position.

WORK OF RELATIVE MOTION.

45. It has been shown that the absolute motion of any system may be resolved into motion of its centre of gravity and motion of its masses relatively to that centre; also, § 42, that the total kinetic energy is separable into two portions, corresponding respectively to those distinct and independent motions, or that

$$\Sigma m \frac{v^2}{2} = \Sigma m \frac{u^2}{2} + \Sigma m \frac{\omega^2}{2}.$$

We may therefore separate the work of these independent motions.

The current co-ordinates of the centre of gravity being denoted by x', y', z', and the relative co-ordinates of m by ξ, η, ζ, the variations of the absolute co-ordinates, § 35, will be

$$dx = dx' + d\xi; \qquad dy = dy' + d\eta; \qquad dz = dz' + d\zeta.$$

If we substitute these values in equations (20) and (26) and suppose the system to start from rest, the constant will be zero, and the work will be

$$\tfrac{1}{2} \Sigma m (u^2 + \omega^2) = \int [X(dx' + d\xi) + Y(dy' + d\eta) + Z(dz' + d\zeta)].$$

But as the relative motions are the same, whether the centre of gravity moves or not, dx', dy', dz', are arbitrary and entirely independent of $d\xi$, $d\eta$, $d\zeta$, as are also u and ω of each other, hence

$$\Sigma m \frac{\omega^2}{2} = \int (X d\xi + Y d\eta + Z d\zeta),$$

or the general theorem of work is applicable to motion relative to the centre of gravity.

ENERGY OF ABSOLUTE AND RELATIVE MOTION.

46. The internal potential energy of any system depends only upon the relative positions of its masses and their mutual actions and reactions. It is, therefore, the same for either the absolute or the relative motion of the system, or

$$\Pi (x, y, z) = \Pi (\xi, \eta, \zeta).$$

If there be no external disturbing forces the motion of the centre of gravity is constant; equation (42) may, therefore, be written thus

$$\Sigma m \frac{v^2}{2} = \Sigma m \frac{\omega^2}{2} + a;$$

and substituting these values in equation (30) we obtain

$$\Pi + \Sigma m \frac{\omega^2}{2} = c.$$

Hence, whenever there are no external forces of power or resistance, the theorem that the sum of the potential and kinetic energy is constant may be applied either for the absolute motion, or for the motion relative to the centre of gravity, and the system is dynamically conservative.

47. For work between the limits (1) and (2) we have found

$$\Sigma \int_1^2 P dp = \tfrac{1}{2} \Sigma m (v_2^2 - v_1^2).$$

This may be divided into the work of the external and that of the internal forces, and be written thus

$$\Sigma \int_1^2 P dp = W.\ int. + W.\ ext.$$

But we have seen that the mutual actions of a system of bodies depend only on the masses and their relative internal positions, so that

$$W.\,int. = \Pi_1 - \Pi_2.$$

Combining these equations we obtain

$$(\Pi_2 - \Pi_1) + \tfrac{1}{2}\Sigma\, m\,(v_2^2 - v_1^2) = W.\,ext.$$

or the variation of the total energy is equal to the work of the external forces.

WORK OF EXPANSION.

48. To obtain an expression of the work done by the pressure of an expanding substance, such as steam or compressed air, acting in all directions with equal force upon equal areas of the enveloping surface, let $dx\,dy$ or ω be an element of that surface, then the outward pressure exerted upon this element by the expansive force p will be $p\omega$; and if it push the resisting surface through the length dz or l, then $dx\,dy\,dz$ or $l\omega$ is the increment of volume dv and pdv is the elementary work. The definite integral

$$\int_1^2 pdv$$

is, therefore, the work done by the expanding substance.

WORK OF HEAT.

49. If we suppose the expansion of any substance to be caused by variation of heat, other changes accompany that of volume.

The increment of heat produces:

1°, a change of invisible molecular motion or vibration, or of temperature, expressed by

$$\tfrac{1}{2}\Sigma\, m\,(\omega_2^2 - \omega_1^2)\,;$$

2°, a change of molecular configuration and consequently of potential energy or latent heat, equal to

$$\Pi_2 - \Pi_1;$$

3°, the change of expansion and external work

$$\int_1^2 pdv.$$

The total variation will, therefore, be

$$\Delta\left(\Sigma\, m\frac{\omega^2}{2} + \Pi + \int pdv\right). \tag{45}$$

This evidently divides into two distinct portions, the invisible change of *internal* energy, kinetic and potential, and the perceptible change of *external* work. Denoting the general integrals of these two portions by U and S, we shall always have for the thermal work of any heated substance

$$\Sigma \int Pdp = U + S. \tag{46}$$

Such unfortunately, in the present state of science, is our ignorance of the constitution of matter, that the function U is generally so hidden as to be indeterminable, though we may often eliminate it. But it is clear that all measurements, however laboriously made of the dynamical action of heat, in which U is neither determined nor eliminated, must be radically defective; and of such there have been unhappily too many. Moreover, equation (46) shows that whenever part of the power is expended in producing thermal vibrations of friction the useful work is thereby diminished.

CALCULATION OF WORK.

50. The expression for mechanical work is

$$u = \Sigma \int Pdp = \int \phi (x)dx = \int ydx,$$

or the same as the geometric formula for quadratures. Whenever, therefore, y is a known function of x, work may be exactly calculated by the method of quadratures.

But if, as is very often the case, Pdp or its equivalent ydx is is not integrable, then its value can only be approximately determined.

As such calculations have often to be made by the professional engineer, we shall conclude this dynamical Introduction by giving the most approved methods; of which there are three: 1° that of trapezoids, 2° that of Thomas Simpson, 3° that of Ponçelet.

METHOD OF TRAPEZOIDS.—Divide the projection of the curve upon the axis of x into equal parts e, and measure the ordinates corresponding to the points of division y_0, y_1, y_2, y_3, *etc.* Then suppose these ordinates to divide the surface into narrow trapezoids, the area of the first is

$$\tfrac{1}{2}e\,(y_0 + y_1),$$

that of the second is

$$\tfrac{1}{2}e\,(y_1 + y_2);$$

and, by summation, the total area is

$$u = e\,[\tfrac{1}{2}(y_0 + y_n) + y_1 + y_2 + y_3 + \ldots + y_{n-1}]. \qquad (47)$$

METHOD OF SIMPSON.—Instead of imagining the curve to be polygonally divided, Simpson applies the fact that, through any three points of a continuous curve, not too remote from each other, a parabola may be drawn, with which that part of the curve may be supposed to sensibly coincide.

Let ACB be the part of the curve, also let $e = ac = cb$. Then for a parabolic segment

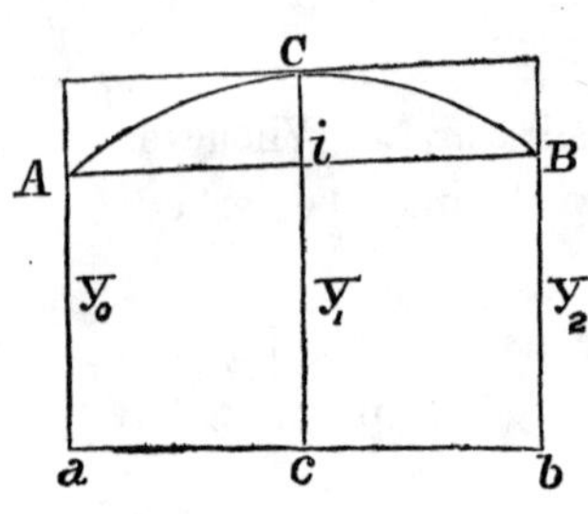

$$AiBC = \tfrac{2}{3}ab \times Ci.$$

Therefore

$$\begin{aligned}\Delta u &= ab\,[ci + \tfrac{2}{3}(Cc - ci)],\\ &= ab\,\tfrac{1}{3}(ci + 2Cc)\,;\end{aligned}$$

or

$$\begin{aligned}\Delta u &= ac\,\tfrac{1}{3}(2ci + 4Cc),\\ &= \tfrac{1}{3}e\,(y_0 + 4y_1 + y_2).\end{aligned}$$

In like manner we find

$$\begin{aligned}\Delta' u &= \tfrac{1}{3}e\,(y_2 + 4y_3 + y_4)\\ \Delta'' u &= \tfrac{1}{3}e\,(y_4 + 4y_5 + y_6).\end{aligned}$$

Hence, by summation

$$\begin{aligned}u = \tfrac{1}{3}e\,[(y_0 + y_n) &+ 4\,(y_1 + y_3 + y_5 + \ldots + y_{n-1})\\ &+ 2\,(y_2 + y_4 + y_6 + \ldots + y_{n-2})]. \qquad (48)\end{aligned}$$

If the curve be reversed, as in the annexed figure, then

$$\begin{aligned}\Delta u &= ab\,[ci - \tfrac{2}{3}(ci - cC)]\\ &= ab\,\tfrac{1}{3}(ci + 2Cc),\end{aligned}$$

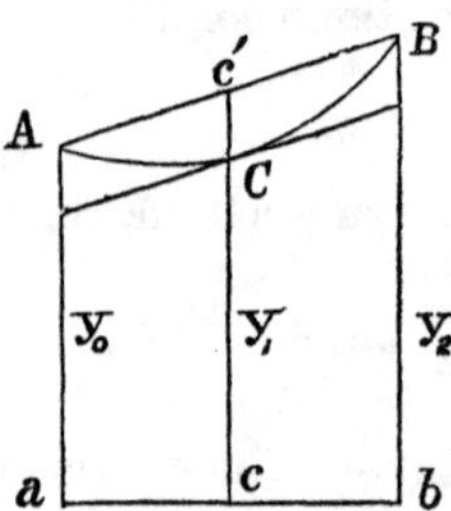

the same as in the former case.

The formula of Simpson is readily obtained algebraically from the equation of the parabola

$$y = \alpha + \beta x + \gamma x^2,$$

which gives

$$\Delta u = \int_0^e y dx = 2\alpha e + 2\beta e^2 + \tfrac{8}{3}\gamma e^3.$$

To determine the factors α, β, γ, the curve gives

$$y_0 = \alpha,$$
$$y_1 = \alpha + \beta e + \gamma e^2,$$
$$y_2 = \alpha + 2\beta e + 4\gamma e^2;$$

making the requisite substitutions and reductions we obtain

$$\Delta u = \tfrac{1}{3}e (y_0 + 4y_1 + y_2),$$

the same as by the geometric construction.

METHOD OF PONCELET.—This method with fewer ordinates, and consequently less labor, gives even a closer approximation than that of the method of Simpson.

It is a modification of the ancient method of exhaustions by inscribed and circumscribed polygons; and the greater the number of subdivisions the more exact will be the result in each of these several methods.

Let AB be the curve, divide the base into an even number of parts, each equal to e, and draw the ordinates, y_0, y_1, y_2, y_3, y_4, etc. The area of the curve will be the mean between those of the circumscribed and incribed trapezoids.

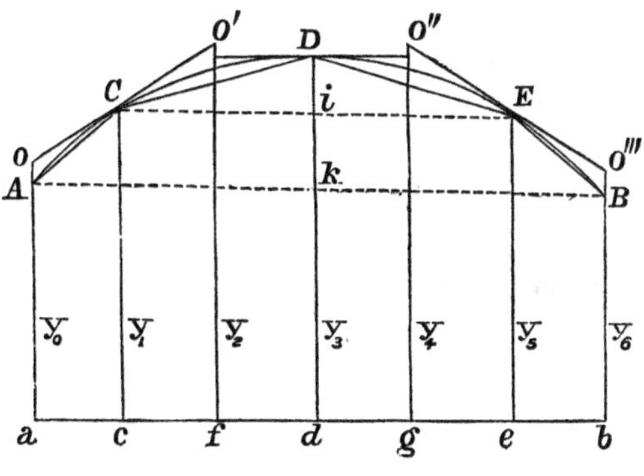

The area of the circumscribed trapezoids is

$$s = 2e\,(Cc + Dd + Ee + etc.) = 2e\,(y_1 + y_3 + y_5 + \ldots + y_{n-1}),$$

and the area of the inscribed trapezoids is

$$s' = \tfrac{1}{2}e\,(Aa + Cc) + e\,(Cc + Dd) + etc. \ldots + \tfrac{1}{2}e\,(Ee + Bb),$$
$$= e\,[\tfrac{1}{2}\,(y_0 + y_1) + y_1 + 2\,(y_3 + y_5 + \ldots + y_{n-3}) + y_{n-1} + \tfrac{1}{2}(y_{n-1} + y_n)];$$

add and subtract $\frac{1}{2}\,(y_1 + y_{n-1})$ and we obtain

$$s' = e\,[2\,(y_1 + y_3 + \ldots + y_{n-1}) + \tfrac{1}{2}\,(y_0 + y_n) - \tfrac{1}{2}\,(y_1 + y_{n-1})],$$

and taking the mean of these areas s and s', we have

$$u = e\,[2\,(y_1 + y_3 + \ldots + y_{n-1}) + \tfrac{1}{4}\,(y_0 + y_n) - \tfrac{1}{4}\,(y_1 + y_{n-1})].$$

The area is, therefore, equal to the product of the interval e by twice the sum of the even ordinates, y_1, y_3, y_5, *etc.*, plus one-fourth of the difference of the sums of the extreme ordinates and those next to the extremes.

As the half sum plus the half difference of s and s' is equal to s the greater, and the half sum less their half difference is equal to s' the less of these two areas, it is evident that the area bounded by the intermediate curve can never differ from the mean value or half sum of s and s', by an amount equal to their half difference. Hence, for this method of Poncelet, the limit of possible error is

$$\frac{s - s'}{2} = \tfrac{1}{4}e\,[(y_1 + y_{n-1}) - (y_0 + y_n)],$$

which the figure shows to be geometrically equal to

$$\tfrac{1}{4}e\,(ki).$$

If this approximation be not sufficient, it may be rendered closer to any required degree by lessening e the interval between the ordinates.

51. To show the comparative accuracy of these several methods, we will apply them to the example of measuring the integrable area between an equilateral hyperbola and one of its asymptotes. The equation of the curve being

$$xy = m,$$

a constant,

$$u = \int_a^b y dx = m \int_a^b \frac{dx}{x} = m \log \frac{b}{a}.$$

If in this expression we put $m = \frac{1}{2}$, $a = 1$, and $b = 7$, then the method of integration gives for u the exact value

$$u = \int_1^7 y dx = \tfrac{1}{2} \log 7.$$

And the several methods give the following comparative results:

Integration, ...	exact,	$u = 0.9730$
Trapezoids, ...	approximation,	$u = 1.0107$
Method of Simpson,	"	$u = 0.9791$
Method of Poncelet,	"	$u = 0.9762$

Hence it appears that the method of trapezoids is in error for this example to the amount of $3\frac{3}{4}$ per cent, that of Simpson to 0.6 per cent, and that of Poncelet to $\frac{1}{3}$ of one per cent. With only the extreme and even ordinates, or little more than half the number required in the method of Simpson, that of Poncelet is, in this example, nearly twice as accurate.

The method of quadratures is always applicable when we have to determine definite integrals of the form

$$u = \int_a^b \phi (x) \, dx;$$

for we may always represent $\phi (x)$ by y, the ordinate of a curve corresponding to the abscissa x; and whatever be the nature of the function, or the quantities x and $\phi(x)$, the value of u can be determined by the area bounded by the curve, its projection on the axis of x, and the ordinates of its consecutive points.

By diminishing the interval e the methods of approximation may be rendered indefinitely closer; and with very few ordinates they usually give results sufficiently accurate for most practical purposes, while integration is rarely possible. These methods are, therefore, of great utility, for they enable us to make readily the calculations required in a vast number of practical questions constantly occurring in engineering, in mechanism, in physics, and in other branches of applied science.

CHAPTER IV.

GENERAL LAWS.

DEFINITION OF TEMPERATURE.

52. Every person is familiar with the sensations to which we apply the adjectives *hot* and *cold* and the word temperature, also with the fact, that when hot and cold bodies act upon each other, heat is lost by the hot and given to the cold, until they become of the same temperature.

But as sensations and adjectives cannot be measured, the thermometer is used to indicate variations of volume which accompany and are functions of the corresponding temperatures of equilibrium into which it puts itself with surrounding bodies.

The ordinary thermometer shows only apparent changes of volume for mercury and glass; and the function which expresses this relative expansion is not even known, other than by aid of an empirical formula which varies with the chemical composition and molecular state of the particular glass employed.

The method used for the graduation of thermometers is based upon the arbitrary assumption, that changes of temperature are proportional to those of volume; which, so far from being true, is generally false; for the law of dilatation of one substance is rarely similar to that of another.

To express this arbitrary assumption algebraically, let v_0 denote the relative volume at 0°, v_1 its volume at 1°, and v_t that at t, then will

$$v_1 - v_0 : 1 :: v_t - v_0 : t \tag{50}$$

4

be the equation which immediately gives for the algebraic definition of the term *temperature*, when indicated by degrees of an ordinary thermometer,

$$t = \frac{v_t - v_0}{v_1 - v_0}. \tag{51}$$

The only substances which are found to obey the law expressed by this definition, with even an approximate accuracy, are a few gases, hydrogen, oxygen, nitrogen, etc., which have hitherto resisted all efforts made to liquefy them by pressure and extreme cold combined. And even for these the law must be considered to apply to their *absolute* dilatation only, or the *relative* dilatation corrected by elimination of that of the glass.

For such gases the observed law of absolute dilatation, usually called the law of Gay Lussac, but which should be named that of Charles, is for changes of volume

$$v = v_0 (1 + \alpha t); \tag{52}$$

in which the coefficient of dilatation α denotes the increment of volume for the cubic unit and for one degree of temperature.

If we observe that, by definition, we have

$$v - v_0 = \alpha v_0 t,$$

it will be evident that equations (51) and (52) are identical.

Permanent gases are also the only substances which obey approximatively the law of Mariotte, that, when air is compressed without change of temperature, the volume varies inversely as the pressure, or

$$pv = p_0 v_0.$$

But if, at the same time, the temperature of the air compressed be elevated to an amount t, then v_0 becomes

$$v_0 (1 + \alpha t),$$

and we get

$$pv = p_0 v_0 (1 + \alpha t); \tag{53}$$

an expression for the laws of Charles and Mariotte combined.

An air supposed to obey exactly those two laws, or their combination expressed by the equation just found, is called a *perfect gas*, or is said to be theoretically in the *perfectly gaseous state*. Of all real gases, hydrogen approximates most nearly to such an hypothetical substance.

As permanent gases are the only substances for which equations (51, 52, and 53) are nearly exact, temperatures should always be measured by air thermometers when accuracy is required. But for ordinary practical and even for many scientific purposes, the indications of mercurial thermometers, between 0° and 100° C., do not differ sufficiently to produce considerable errors. And, when precision is requisite, corrections may be applied to reduce degrees observed with a mercurial thermometer to their corresponding values indicated by the expansion of air; for which purpose Regnault has furnished the requisite data, and even a table of equivalent indications extended to 350° C. (See Mem. de l'Inst., t. xxi, p. 239.) In the theoretical discussion of thermodynamic phenomena, temperatures are, therefore, always to be supposed to be those given by the absolute dilatation of air.

QUANTITIES OF HEAT.

53. There is an obvious distinction between temperatures and quantities of heat. To heat a cubic foot of water, weighing 1000 ounces avoirdupois, through a given range of temperature, it would evidently be necessary to consume 1000 times the amount of fuel requisite for one ounce. A large block of ice would require more heat to melt it than a small one. Also a ton and an ounce of red-hot iron may be each of the same temperature, though there would manifestly be far more of what is called heat in the ton than in the ounce.

Yet proper attention was not paid to this simple but important distinction until it was shown by Dr. Black, at the middle of the last century, that of the large quantity of heat required to melt solids, or to evaporate liquids, none whatever is indicated by a thermometer. Hence he gave the name *latent heat* to that which thus causes such changes without elevation of temperature.

When bodies put themselves into equilibrium of temperature with each other, these changes are attended with transfers of heat absorbed by one and given off by another, to which the terms *specific heat* and *calorific capacity* are generally applied.

In the investigation of all such thermodynamic phenomena, quantities of heat must be measured; and for these measurements a standard unit is necessary. The *calorie*, or unit by weight of water at 0° raised to 1° C., is conventionally the *thermal unit* employed, and to this common measure all *thermal quantities* may be reduced.

It was long imagined that the quantity of heat requisite to raise a given body from any temperature t to the consecutive degree $(t+1)°$ is constant, whatever may be the value of t; but this has been shown to be untrue. Hence it is necessary to fix the temperatures 0° and 1°, for the standard unit. Generally, if we suppose the temperature of any body to be t, and that it takes an increment dt, in consequence of the reception of an amount of heat dq, the specific heat of the body, at that temperature t, is

$$\frac{dq}{dt} = f(t)\,; \tag{54}$$

a function of the temperature which varies with the nature of the substance.

The terms quantity of heat, calorie, latent and specific heat, capacity for heat, calorimetry, etc., are due to the material hypothesis; according to which heat was a subtle, indestructible substance, called *caloric*, combining with or separating from other matter. But it is perfectly easy to think of a quantity of heat as energy, or as an amount of vibratory motion; which may be transferred from one body to another, and thus be lost or gained, communicated or received. Hence, those terms, which have long been the familiar names for certain observed facts, may still be used in the new dynamical theory of heat, without confusion of thought; and indeed, it would be difficult, even if it were desirable, to find for them equivalent words.

GENERAL FORMULAS.

54. To determine the quantity of heat corresponding to any given amount of energy or mechanical work, equation (46) and the law of Joule give

$$EQ = \Sigma \int Pdp = U + S; \tag{55}$$

in which E denotes the mechanical equivalent of heat, and is called Joule's coefficient.

And if we denote by A the reciprocal of E, or the *heat equivalent* of work, then

$$Q = A\,(U + S). \tag{56}$$

If the sensible or external work S be that done by the pressure of an expanding substance, such as steam or air, then

$$S = \int pdv,$$

and equation (56) may be put under the form

$$Q = AU + A \int pdv. \tag{57}$$

Taking this between the states or limits (1) and (2), we obtain

$$Q_2 - Q_1 = A\,(U_2 - U_1) + A\int_1^2 pdv. \tag{58}$$

If the body or system pass from the state (1), by any intermediate cycle of thermodynamic changes, back to the same state, or configuration (1), then will

$$U_1 - U_1 = W.\ int. = \Pi_1 - \Pi_1 = 0, \tag{59}$$

and equation (57) reduces to

$$Q = A\int_1^1 pdv. \tag{60}$$

This result is of great importance, for it shows how, in thermal investigations, we can eliminate the internal energy U; which is generally inaccessible to experimental determination. Consequently, if it were impracticable to eliminate U, we might despair of being able to make any considerable progress in this branch of science. The necessity and the mode of such an elimination were first indicated by Sadi Carnot, who drew attention to the truth, that thermal energy can be continuously converted into external useful work only when the system periodically returns to the same initial state, or configuration, and the variations of the internal energy of the system, or of its potential Π, reduce to zero.

THERMODYNAMIC FUNCTIONS.

55. The thermodynamic state of any body is a function of the action upon it of external forces, of the temperature, and of its specific volume or reciprocal of its density. As the external forces may vary in any manner whatever, the problem of determining that state is evidently too general for solution. It is, therefore, simplified by assuming the external forces to be only those of a normal pressure acting uniformly upon the containing surface, and also by supposing the density and temperature constant throughout the entire mass. Less simple cases are referred to the theory of elasticity of bodies. But even when thus limited and simplified, the problem is often incapable of solution.

As the dynamical condition of any system is determined by its total energy, kinetic and potential, if we denote these by Υ and Π, the normal pressure, temperature, and specific volume, or any other dependent variables, such as the conductivity, radiation, index of refraction, etc., are functions of Υ and Π; and from any three equations,

$$p = f(\Upsilon\, \Pi), \qquad v = f'(\Upsilon\, \Pi), \qquad t = f''(\Upsilon\, \Pi),$$

the variables Υ and Π may be eliminated, leaving only one equation,

$$\phi\,(pvt) = 0. \tag{61}$$

Only for the permanent gases has it been found possible to determine the form of this function with sufficient accuracy. The combined law of Charles and Mariotte (53) is, as we have stated, a limit to which they approach, more or less closely, in their thermodynamic changes; hydrogen obeying that law very nearly, while carbonic acid and other liquefiable gases or vapours depart sensibly from it.

PARTIAL DIFFERENTIAL EQUATIONS OF TRANSFORMATION.

56. We may regard the thermodynamic state of any body or system as a determinate function of three variables, the temperature t, the specific volume v, and the normal pressure p; it is expressed by the equation (61), just found. Also these variables p, v, t, may be considered as co-ordinates of a geometric surface representing the function.

If for any assumed value of t we find the corresponding values of p and v, this is equivalent to the determination of the form of a section of the thermodynamic surface perpendicular to the axis of t. Similarly, we may find any number of sections, each perpendicular to the same axis; and these sections would evidently determine the surface. But as sections may be thus taken perpendicularly to each of the three axes, the investigation corresponding to this geometrical analysis by sections may be made in three distinct ways; in each of which two of the co-ordinates, p, v, t, are taken as variables, while the third is an arbitrary constant. It is clear that the results thus obtained must all accord, as they are related to each other by the common function (61), which is represented by the same geometric surface, and expresses the states, or transformations, of the body.

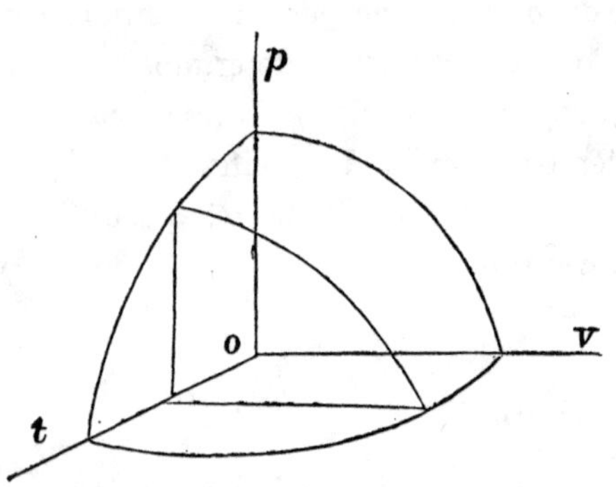

57. Let v and t be the variables for any constant value of p. If now the temperature become $t + dt$, and the specific volume $v + dv$, then will

$$dQ = \frac{dQ}{dt}dt + \frac{dQ}{dv}dv$$

be the variation of the quantity of heat for such a change. Substituting letters for the partial derivatives

$$c = \frac{dQ}{dt}, \qquad l = \frac{dQ}{dv}, \tag{62}$$

the equation just found becomes

$$dQ = cdt + ldv. \tag{63}$$

The partial derivative c is usually called the *specific heat of constant volume*, it being the quantity of heat requisite to produce a given change of temperature without any variation of volume. And, analogously, the coefficient l bears the name of the *latent heat of dilatation.*

For any indefinite change we have

$$Q = \int (cdt + ldv). \tag{64}$$

But to integrate the second member of this equation, it is necessary to know the function (61) for the particular body or system. And as that function is not known, even approximately, except for permanent gases, the integration is rarely possible.

Moreover, such is the impracticability of confining any solid or liquid body when heated or chilled, and of preventing for compressible gases the loss or gain of heat by conduction, that the coefficient c cannot be ascertained experimentally; while for the quantity l, though attainable, we possess few observations.

58. Experimental investigations have been mostly of specific heat, or calorific capacity, under constant pressure. Taking, therefore, p and t for variables, we have

$$dQ = \frac{dQ}{dt}dt + \frac{dQ}{dp}dp;$$

and denoting the *specific heat of constant pressure* by c',

$$c' = \frac{dQ}{dt};$$

if p vary independently of t, we have hdp for the quantity of heat necessary to this change; the coefficient h being analogous to c, l, c', but without a conventional name. Thus, we obtain

$$dQ = c'dt + hdp, \tag{65}$$

for the elementary quantity of heat due to this transformation.

Between (63) and (65) there exists a relation determined by the function (61); seeking from which the value dv, or of

$$dv = \frac{dv}{dt}dt + \frac{dv}{dp}dp,$$

we get, by substitution,

$$c'dt + hdp = cdt + l\left(\frac{dv}{dt}dt + \frac{dv}{dp}dp\right);$$

and equating coefficients of like quantities,

$$c' = c + l\frac{dv}{dt},$$

$$h = l\frac{dv}{dp}. \tag{66}$$

59. Finally, if p and v be taken as the variables, we similarly have for an elementary variation of heat,

$$dQ = Mdv + Ndp; \tag{67}$$

M and N being nameless coefficients analogous to c, l, c', h; and for which the function (61) gives the following relations,

$$dt = \frac{dt}{dv}dv + \frac{dt}{dp}dp,$$

consequently,

$$Mdv + Ndp = ldv + cdt$$

$$= ldv + c\left(\frac{dt}{dv}dv + \frac{dt}{dp}dp\right).$$

From which we readily deduce

$$M = l + c\frac{dt}{dv} = c'\frac{dt}{dv},$$

$$N = c\frac{dt}{dp}. \tag{68}$$

60. It is often desirable and even necessary, in the investigation of thermal phenomena, to take for independent variables, instead of the pressure, temperature and specific volume, other data, such as the conductivity, the emissive or absorbent power, the index of refraction, etc., of the substance under consideration.

Denoting, therefore, by x and y any two such variables upon which the thermal state of the body depends, or varying with it in any manner whatever, we shall have for the elementary quantity of heat corresponding to their variation,

$$dQ = mdx + ndy\,; \tag{69}$$

and if dv and dt be the corresponding variations of volume and temperature, then

$$mdx + ndy = cdt + ldv.$$

Consequently,

$$mdx + ndy = c\left(\frac{dt}{dx}dx + \frac{dt}{dy}dy\right) + l\left(\frac{dv}{dx}dx + \frac{dv}{dy}dy\right),$$

and, therefore, we have always

$$\left.\begin{aligned} m &= c\frac{dt}{dx} + l\frac{dv}{dx}, \\ n &= c\frac{dt}{dy} + l\frac{dv}{dy}; \end{aligned}\right\} \tag{70}$$

general formulas, of which (66) and (68) are only particular values.

RELATIONS OF THE PARTIAL DIFFERENTIAL COEFFICIENTS.

61. For any elementary quantity of heat equation (57) gives the general value

$$dQ = AdU + Apdv. \tag{71}$$

But as U is a function of p and v,

$$dU = \frac{dU}{dv}dv + \frac{dU}{dp}dp,$$

and

$$dQ = A\left(\frac{dU}{dv} + p\right)dv + A\frac{dU}{dp}dp.$$

Now, by equation (67),

$$dQ = Mdv + Ndp,$$

consequently,

$$M = A\left(\frac{dU}{dv} + p\right),$$

$$N = A\frac{dU}{dp}.$$

From which,

$$\frac{dM}{dp} = A\frac{d^2U}{dv \cdot dp} + A,$$

$$\frac{dN}{dv} = A\frac{d^2U}{dp \cdot dv};$$

and, therefore,

$$\frac{dM}{dp} - \frac{dN}{dv} = A. \tag{72}$$

Let now v and t be the variables of which U and p are functions, then

$$dU = \frac{dU}{dt}dt + \frac{dU}{dv}dv,$$

$$dQ = A\frac{dU}{dt}dt + A\left(\frac{dU}{dv} + p\right)dv.$$

But equation (63) gives

$$dQ = cdt + ldv;$$

therefore,

$$l = A\left(\frac{dU}{dv} + p\right),$$

$$c = A\frac{dU}{dt};$$

which give

$$\frac{dl}{dt} = A\frac{d^2U}{dv\,dt} + A\frac{dp}{dt},$$

$$\frac{dc}{dv} = A\frac{d^2U}{dt\,dv};$$

consequently, there always exists the relation

$$\frac{dl}{dt} - \frac{dc}{dv} = A\frac{dp}{dt}. \tag{73}$$

Lastly, if we take p and t for variables, of which U and v are functions, then

$$dU = \frac{dU}{dt}dt + \frac{dU}{dp}dp,$$

$$dv = \frac{dv}{dt}dt + \frac{dv}{dp}dp,$$

$$dQ = A\left(\frac{dU}{dt} + p\frac{dv}{dt}\right)dt + A\left(\frac{dU}{dp} + p\frac{dv}{dp}\right)dp.$$

But equation (65) gives

$$c' = A\left(\frac{dU}{dt} + p\frac{dv}{dt}\right);$$

$$h = A\left(\frac{dU}{dp} + p\frac{dv}{dp}\right);$$

from which,

$$\frac{dc'}{dp} = A\left(\frac{d^2U}{dt\,dp} + p\frac{d^2v}{dt\cdot dp} + \frac{dv}{dt}\right),$$

$$\frac{dh}{dt} = A\left(\frac{d^2U}{dp\,dt} + p\frac{d^2v}{dp\cdot dt}\right);$$

and consequently

$$\frac{dc'}{dp} - \frac{dh}{dt} = A\frac{dv}{dt}. \qquad (74)$$

There relations (72), (73), (74), between the coefficients c. c', l, etc., which we owe to Clausius, show that the partial differential equations of transformation (63), (65), (67), are not directly integrable; for the criterion of integrability

$$\frac{dm}{dy} = \frac{dn}{dx},$$

showing an expression of the form

$$mdx = ndy$$

to be an exact differential, is not fulfilled in either case. Yet there always exists for such an expression an infinite number of factors such that, if multiplied by any one of them, the expression becomes an exact differential capable of integration. This is usually proved in elementary treatises on the integral calculus, but for convenient use we give a brief demonstration.

FACTORS OF INTEGRABILITY.

62. For any constant value of one of the co-ordinates or variables, such as t, or for any given section of the corresponding geometric surface cut perpendicularly to an axis, the general function (61) reduces to one of the two remaining co-ordinates, or variables, and an arbitrary constant,

$$\phi\,(p, v, c) = 0.$$

By differentiation,

$$d\,\phi = \frac{d\phi}{dv}\,dv + \frac{d\phi}{dp}\,dp = 0;$$

and it is evident that, if we compare this with

$$Mdv + Ndp,$$

we must have

$$\frac{dM}{dp} = \frac{d^2\phi}{dp\,dv} = \frac{dN}{dv},$$

as the criterion of integrability, whenever the expression is an exact differential of which the function ϕ is the integral.

We have seen that the partial differential equations of transformation by heat, such as

$$dQ = Mdv + Ndp,$$

do not fulfil the criterion of integrability just demonstrated. Let us, therefore, put this equation under the equivalent form

$$\frac{dp}{dv} + u = 0,$$

in which u will be a function of p and v. The general integral of this equation is a function of p and v and of an arbitrary constant. Let this integral, resolved with reference to the constant, be

$$\phi\,(p, v, c) = 0.$$

Then, by differentiation, we get

$$\frac{d\phi}{dv} + \frac{d\phi}{dp} \cdot \frac{dp}{dv} = 0, \quad \text{or} \quad \frac{dp}{dv} + \frac{d\phi}{dv} \cdot \frac{dp}{d\phi} = 0,$$

an equation from which the arbitrary constant has been eliminated; and which should, therefore, be identical with that proposed. Consequently,

$$\frac{dp}{dv} + u = \frac{dp}{dv} + \frac{d\phi}{dv} \cdot \frac{dp}{d\phi},$$

or multiplying both sides by the same factor,

$$\frac{d\phi}{dp}\left(\frac{dp}{dv} + u\right) = \frac{d\phi}{dv} + \frac{d\phi}{dp}\cdot\frac{dp}{dv}.$$

But the second member is simply the first derivative of the function ϕ; such is also the case, therefore, with the proposed equation

$$\frac{dp}{dv} + u = 0,$$

when multiplied by the requisite factor.

Hence, denoting by λ the reciprocal of the factor of integrability, we have

$$\frac{dQ}{\lambda} = \frac{M}{\lambda} dv + \frac{N}{\lambda} dp = d\phi; \tag{75}$$

and there is, therefore, a factor $\frac{1}{\lambda}$ which renders a differential of the proposed form exact and integrable.

The number of such factors is also infinite, for let z be one suitable to render exact

$$z\,Mdv + z\,Ndp = d\omega,$$

then will

$$z\,\psi\omega\,(Mdv + Ndp) = \psi\omega\,d\omega$$

be an exact differential, and as $\psi\omega$ is an arbitrary function, it may evidently have an infinite number of values.

LAWS OF THE PERFECTLY GASEOUS STATE.

63. The consideration of gases assumed to obey exactly the law of Charles and Mariotte,

$$pv = p_0 v_0 (1 + \alpha t),$$

and therefore said to be theoretically perfect, is very important; for thus we are enabled to discuss problems in thermodynamics in a manner precisely analogous to that which has been so advantageously used in mechanics for the lever, the pulley, the pendulum, the parabolic motion of projectiles and many other problems, in which the effects of friction, resistance, and other disturbing actions are provisionally disregarded, for the purpose of obtaining simpler approximate solutions, which need only small corrections to be applied to them to become close expressions of the real facts of nature. This course is also followed in astronomy; the imaginary elliptical orbits of planets being only first approximations, or hypotheses, requiring to be corrected for perturbations. What those fictitious elliptical orbits are to astronomy, what a frictionless machine in vacuo, or a simple pendulum, is to mechanics, such in the study of heat is a theoretically perfect gas, an important simplification giving for difficult problems approximate solutions; which either differ insensibly from exact solutions, or require only slight corrections to render them accurate enough for all purposes. In fact, they determine the first terms of a convergent series, whose remaining terms either are inappreciable, or else constitute definite residual phenomena for extended investigation.

Thus they establish positive relations, or laws, which must be included in any future more advanced state of knowledge, and indicate the path of research and discovery.

Moreover, they fully explain hot-air engines, such as that of Ericsson, and enable us to compare them correctly with the steam-engine.

64. We shall also assume as laws, or as postulates, for theoretically perfect gases, obeying the law of Charles and Mariotte, the following experimental inductions:

1°, the second law of Joule, that the internal energy of permanent gases is a function of the temperature only, and therefore does not vary with the density or specific volume;

2°, the law of Regnault, that the specific heat of constant pressure of any permanent gas is independent of its temperature and density;

3°, the experimental law, that the product of the density by the specific heat of constant pressure is the same constant for all permanent gases.

The last two of these laws have been satisfactorily established by Regnault; who has also determined the deviation of atmospheric air, oxygen, hydrogen, and carbonic acid from the combined law of Charles and Mariotte. The second law of Joule has been verified both by Regnault and by Sir W. Thomson. And the following important consequences are readily deduced.

By differentiating U in equation (71) as a function of v and t, we get

$$dQ = A\frac{dU}{dt}dt + A\left(\frac{dU}{dv} + p\right)dv.$$

But by the second law of Joule, U is a function of t only for permanent gases, and does not vary with v, hence

$$\frac{dU}{dv} = 0\,; \tag{76}$$

and the equation just found reduces to

$$dQ = A\frac{dU}{dt}dt + Apdv. \tag{77}$$

If now we compare this with the equation of transformation, when v and t are independent variables,

$$dQ = cdt + ldv,$$

we see that, for perfect gases,

$$c = A\frac{dU}{dt}\,; \qquad \text{and} \qquad l = Ap. \tag{78}$$

Hence the specific heat of constant volume of any perfect gas can only be a function of its temperature. And the latent heat of dilatation is proportional to the pressure.

65. We have found, see § 61, for the specific heat of constant pressure, the general value

$$c' = A\left(\frac{dU}{dt} + p\frac{dv}{dt}\right);$$

but the law of Charles and Mariotte gives

$$p\frac{dv}{dt} = \alpha p_0 v_0;$$

therefore

$$c' = A\left(\frac{dU}{dt} + \alpha p_0 v_0\right); \qquad (79)$$

consequently, the specific heat of constant pressure can vary only with temperature. But, by the postulate or experimental law of Regnault, it is independent of the temperature; and, therefore, is a constant for each gas.

66. The specific heat of constant volume must also be a constant for each gas; for by subtraction,

$$c' - c = A\,\alpha p_0 v_0; \qquad (80)$$

the second member of which is constant for the same substance.

And if we integrate the expression

$$c = A\frac{dU}{dt},$$

under the hypothesis that c is constant for any particular gas, then we will obtain

$$AU = ct, \qquad \text{or} \qquad U = Ect; \qquad (81)$$

the algebraic expression for the second law of Joule. The constant of integration being zero.

67. If now we divide both sides of equation (80) by v_0, or which is the same thing, multiply by δ, the density or specific gravity, we have

$$\frac{c' - c}{v_0} = (c' - c)\,\delta = A\,\alpha\,p_0; \tag{82}$$

and as the second member is the same constant for all the permanent gases, so must the first be also.

But by the third postulate, or law of Regnault, the ratio

$$\frac{c'}{v_0} = c'd \tag{83}$$

is the same constant for all permanent gases; consequently,

$$\frac{c}{v_0} = cd \tag{84}$$

must also be the same constant for them all.

JOULE'S COEFFICIENT THEORETICALLY DETERMINED.

68. From equation (82) we obtain the theoretical formula

$$E = \frac{1}{A} = \frac{\alpha\,p_0 v_0}{c' - c}, \tag{85}$$

which enables us to calculate the mechanical equivalent of Joule, and compare this value with that determined by his experiments on friction.

For hydrogen, oxygen, and atmospheric air, all the quantities in the second member of (85), except c, have been experimentally determined with great precision by Regnault. The value of c

is less accurately known, for it is not possible to obtain it by direct experiment; indirectly, however, it has been deduced from the velocity of sound. If the acoustical value of c be employed, with those of Regnault for c', α, v_0, the following values of E, the equivalent of Joule, are given by the permanent gases,

Hydrogen	425.3
Oxygen	425.7
Atmospheric air	426.

Of these results, that given by hydrogen, which approaches very nearly to the limit of the perfectly gaseous state, is the most probable.

The accordance of these theoretically computed values, based upon entirely independent data, with the coefficient obtained experimentally from friction by Joule, is truly remarkable, and fully verifies the accuracy of his work.

DETERMINATION OF THE FACTOR OF INTEGRABILITY.

69. We have proved that there exists always a factor, which can render integrable a partial differential equation of thermodynamic change. That factor may be easily found for perfect gases; for writing the law of Charles and Mariotte under the form

$$pv - \alpha p_0 v_0 (a + t) = 0, \tag{86}$$

in which a is the reciprocal of α, and substituting, in

$$dQ = cdt + ldv,$$

for l its value Ap, we have

$$dQ = cdt + Apdv; \tag{87}$$

whence we obtain, by replacing p by its value given in (86), the expression

$$dQ = cdt + A\,\alpha\,p_0v_0\,(a + t)\,\frac{dv}{v};$$

and consequently,

$$\frac{dQ}{a+t} = \frac{cdt}{a+t} + A\,\alpha\,p_0v_0\,\frac{dv}{v}; \qquad (88)$$

the second member of which is evidently an exact differential; for c is a function of t only, and the variables are separated. Calling it $d\,(\phi)$, we have

$$\frac{dQ}{a+t} = d\,(\phi). \qquad (89)$$

One of the values for λ, the reciprocal of the factor of integrability, is, therefore,

$$\lambda = a + t.$$

If p and t be the independent variables, then equation (65) gives

$$dQ = c'dt + hdp,$$

and substituting for h its value, given by equations (66) and (78), or

$$h = l\,\frac{dv}{dp} = Ap\,\frac{dv}{dp},$$

we obtain

$$dQ = c'dt + Ap\,\frac{dv}{dp}\,dp.$$

But equation (86) gives

$$p\,\frac{dv}{dp} = -\,v = -\,\frac{\alpha\,p_0v_0}{p}\,(a + t),$$

therefore

$$h = -\,Av; \qquad (90)$$

and

$$dQ = c'dt - Avdp; \qquad (91)$$

consequently,

$$dQ = c'dt - A\,\alpha p_0 v_0\,(a + t)\,\frac{dp}{p},$$

and

$$\frac{dQ}{a + t} = \frac{c'}{a + t}\,dt - A\,\alpha p_0 v_0\,\frac{dp}{p}. \tag{92}$$

Hence, as c' is a function of t only, the variables are separated and the second member is the exact differential of a function of p and t, so that again

$$\frac{dQ}{\lambda} = \frac{dQ}{a + t} = d\,(\phi).$$

Lastly, if we assume p and v for the independent variables, then will

$$dQ = Mdv + Ndp\,;$$

and substituting in this for M and N the values given by (68), it becomes

$$dQ = ldv + c\left(\frac{dt}{dv}\,dv + \frac{dt}{dp}\,dp\right);$$

but the quantities within brackets are the partial differentials of t as a function of p and v, and their sum is its total differential dt, hence this equation reduces to

$$dQ = cdt + ldv\,;$$

which is the same as equation (87) and is rendered integrable by the same factor,

$$\lambda = a + t.$$

Hence, for the perfectly gaseous state, there is a factor λ, equal to the temperature t plus a constant a, which renders integrable the equations of thermal transformation ; and as the constant a is the reciprocal of α, the coefficient of dilatation in the law of Charles and Mariotte, the factor λ is the same for all perfect gases. We will, hereafter, show that this factor λ is also the same for all substances whatever.

ABSOLUTE TEMPERATURES AND AN ABSOLUTE ZERO OF HEAT.

70. As heat is energy of motion and all motion may be reduced to repose, it necessarily follows, whatever be the form of those hidden molecular motions which render bodies hot, that they can come to rest, or end in an *absolute zero* of heat and temperature. As silence is complete negation of sound, and darkness of light, so is this thermal zero utter privation of heat. Moreover, from these dynamical views, it follows that negative absolute temperatures cannot exist.

If in equation (87) we suppose that no external work is done, then pdv is zero, and

$$dQ = cdt + Apdv$$

reduces to

$$dQ = cdt;$$

which, by integration between limits, becomes

$$Q - Q_0 = c\,(t - t_0), \tag{93}$$

provided that c is constant. Now, the researches of Regnault prove that, though c is variable for vapours, liquids and solids, it may be considered a constant for each permanent gas. Consequently, the law expressed by equation (93), that variations of temperature are proportional to the corresponding changes in quantity of heat, is true only for perfect, or permanent, gases expanding without external work. Only air thermometers, therefore, can be used for determining with precision corresponding changes of temperature and quantity of heat.

But as each observation made with an air thermometer is a delicate and difficult experiment, and as the function c is determinate for every substance, mercurial thermometers may be graduated by comparison with air thermometers, and thus be rendered accurate. In fact, the great requisite for thermometers is that their indications be comparable with each other, and with those of air thermometers; which alone can serve as standards to measure temperatures proportionally to variations of heat or energy. From equation (81) we have

$$U - U_0 = Ec\,(t - t_0)\,; \tag{94}$$

which shows that, when pdv is zero, the variation of internal energy of a perfect gas is proportional to that of the temperature indicated by an air thermometer.

71. To determine absolute temperatures, suppose, in the equation

$$pv = \alpha p_0 v_0\,(a + t),$$

the expansive force p to become zero, so that the gas exerts no tension and can do no work; then, whatever may be the value of v, we shall have

$$a + t = \frac{1}{\alpha} + t = 0.$$

But according to the measurements of Regnault, the value of a is 273 very nearly. Consequently,

$$a + t = 273 + t = 0$$

is the equation which determines the ordinary centigrade temperature t corresponding to the *absolute zero* of heat. Its value is evidently $-273°$ C., which is nearly equal to $-460°$ Fahrenheit.

Denoting absolute temperatures by the Greek letter τ, and those of the ordinary thermometer by t, we shall have generally

$$\tau = a + t. \tag{95}$$

Hence, to convert centigrade temperatures of the air thermometer into absolute temperatures, proportional to variations of energy or heat, we have only to add to the observed indications the constant number a or 273° C. This is evidently a mere change of origin of co-ordinates.

Absolute temperatures can be as readily expressed in degrees Fahrenheit, or Reaumur; the algebraic formula being the same, the only differences are in the arithmetical values of a and t required by their respective scales.

Differences of temperature give always

$$\tau_2 - \tau_1 = t_2 - t_1;$$

or they are equal for both absolute and ordinary temperatures; consequently dt is always the same as $d\tau$. For many purposes, only changes, or differences, of temperature need to be expressed; and the smaller numbers of the centigrade thermometer are often more convenient than when increased by 273 to reduce them to the absolute scale. But thermodynamic formulas are generally simplified by substituting τ, the absolute temperature, in place of $a + t$, the ordinary centigrade or Fahrenheit indication.

As both the factor λ and the absolute temperature τ have been shown to be equal to the same quantity, so that

$$\lambda = a + t = \tau, \tag{96}$$

it follows that *the absolute temperature τ is the factor of integrability λ for perfect gases.* This result will be shown to be general, or applicable to all bodies.

CHAPTER V.

AIRS AND VAPOURS.

LAWS OF CHARLES AND MARIOTTE.

72. To deduce the thermodynamic laws of elastic fluids, such as air, from the general differential equations of transformation and energy, we have supposed gases to obey exactly certain experimental laws, § 64, among which are the law of Mariotte,

$$pv = p_0 v_0, \tag{96}$$

and that of Charles,

$$v = v_0 (1 + \alpha t), \tag{97}$$

usually called the law of Gay Lussac.

The results were presented to your attention as close approximations to the real phenomena; and the degree of the approximation was compared to that of elliptical planetary orbits, requiring only small corrections for perturbations.

To justify this comparison, and let you judge of the probable exactness of conclusions thus theoretically demonstrated, it is well to present briefly the results of the most reliable experimental investigations.

LAW OF MARIOTTE.

73. This fundamental law, sometimes called that of Boyle, was for nearly two centuries believed to be exact for all gases.

Boyle and Muschenbroeck, however, had each been experimentally led to believe the compression of air less for high pressures than the law assumes; but Sulzer, in 1753, published experiments showing it to be greater. Robison, imagining the results of Sulzer to be due to condensation of hygrometric moisture, experimented upon moist air, air dried with caustic lime, and air containing vapour of camphor, but found differences from the law of Mariotte even greater than those of Sulzer. The following are his observations:

DENSITY.	ELASTIC FORCE.
1.000	1.000
2.000	1.957
3.000	2.848
4.000	3.737
5.500	4.930
6.000	5.342
7.620	6.490

Without doubt, these very inaccurate results were caused by moisture, or other sources of error.

In 1826, experiments made with improved instruments were published by Oersted and Swendsen, which seemed to show, though not distinctly, a compression for air slightly greater than that of the law of Mariotte. But for sulphurous acid they found the compressibility far greater than that of air.

It was discovered by Faraday, that chlorine, sulphurous acid and many other gases are liquefied by pressure and cold. And Despretz confirmed the experiments of Oersted and Swendsen upon sulphurous acid; which he also extended to sulphuretted hydrogen, ammonia and cyanogen; proving them more compressible than atmospheric air. But he found that hydrogen and air do not differ for pressures below 15 atmospheres; though they appeared to differ above 20 atmospheres.

All doubt concerning the exactness of the law of Mariotte for atmospheric air seemed finally to be completely dispelled by the experiments of Dulong and Arago, made by request of the

Academy of Sciences and published in 1829; their observations differed from the theoretical values in no instance more than 0.01, for pressures from 1 to 27 atmospheres; and the differences seemed purely accidental, following no law. Hence, they concluded that the law of Mariotte is perfectly exact for air and permanent gases, for all pressures below 27 atmospheres, and most probably also for those above that limit.

It was the wish and intention of Arago and Dulong to have continued and extended their researches, but this was prevented. Pouillet, therefore, made experiments, the results of which were briefly given in his *Elémens de Physique,* and are as follows: oxygen, nitrogen, hydrogen, and carbonic oxide follow the same law as atmospheric air up to 100 atmospheres; sulphurous acid, ammonia, carbonic acid and the protoxide of nitrogen are much more compressible than atmospheric air; protocarburetted and bicarburetted hydrogen, which at 10° C. are not liquefied by a pressure of 100 atmospheres, are sensibly more compressible than air.

74. Such was the state of knowledge upon this subject before it was investigated by Regnault, clearly showing liquefiable gases to be more compressible than air, but proving the law of Mariotte for the compression of permanent gases to be an approximation so close as to have eluded all detection of difference by such skilful observers as Dulong, Arago and Pouillet. Consequently, our comparison with planetary orbits is fully justified; and any small differences which exist may be properly considered perturbations.

We shall not attempt to give a complete account of the researches of Regnault; made with that extreme accuracy for which he was distinguished, they can only be appreciated by reading the original memoir published, in 1847, in the transactions of the Institute, t. xxi, p. 329.

The following abbreviated table presents his results for several gases:

DENSITY.	ELASTIC FORCE OF PERMANENT GASES.					
	Hydrogen.		*Nitrogen.*		*Atmospheric Air.*	
2	2.0008	1.0004	1.9995	0.9992	1.9975	0.998782
4	4.0061	1.0015	3.9918	0.9979	3.9860	0.996490
8	8.0339	1.0042	7.9641	0.9955	7.9457	0.993212
16	16.1616	1.0101	15.8597	0.9912	15.8045	0.987780

The third, fifth, and seventh vertical columns give the ratios of the pressure, or elastic force, to the corresponding density. For atmospheric air, the maximum difference between the observed and theoretical results, or the deviation from the law of Mariotte, sensibly exceeds one per cent, and amounts to 0.0122 only under a compression of sixteen times the original density. Hydrogen presents the singular phenomenon, that the ratio of the elasticity to the density increases with the pressure, while it diminishes for air and nitrogen.

For carbonic acid, one of the readily liquefiable gases, Regnault obtained the following results:

DENSITY.	PRESSURE.	RATIO.	DENSITY.	PRESSURE.	RATIO.
2	1.9829	0.9915	8	7.5193	0.9399
4	3.8969	0.9742	16	13.9267	0.8704

Hence it appears that carbonic acid deviates to the amount of 0.13 from the law of Mariotte for a pressure of sixteen atmospheres, or more than three-fourths of one per cent for each atmosphere. Numerous observations for each of the above cases are recorded in the original memoir. They were all graphically represented by curves, and for each curve Regnault has given a formula for the computation of the elastic force for any density, thus enabling us to correct for deviations from the law of Mariotte, whenever great precision is required. But for most practical purposes, such corrections are not necessary.

To get the formula for atmospheric air, let m denote the density corresponding to any elastic force r, then will

$$\frac{v_0}{v} = m, \qquad \frac{p}{p_0} = r;$$

and for the law of Mariotte, their ratio is

$$\frac{pv}{p_0 v_0} = \frac{r}{m} = 1;$$

but if that law be not exact for air, their ratio may be considered a parabolic function of $(m - 1)$ and

$$\frac{r}{m} = 1 + A\,(m - 1) + B\,(m - 1)^2. \qquad (93)$$

To determine the coefficients A and B, the values of m, or densities 8 and 16, give

$$\frac{r}{m} = 0.993212, \qquad \frac{r}{m} = 0.98778;$$

from which Regnault deduces

$$\log A = \bar{3}.0435120,$$
$$\log B = \bar{5}.2873750.$$

Subsequently, Regnault obtained for other gases, when r is sensibly equal to 2, the following values of m divided by r, or of the ratio of the density to the corresponding pressure or elastic force:

Atmospheric air . .	1.00215	Acid hydrochloric . .	1.00925
Nitrogen binoxide . .	1.00285	" sulphohydric . .	1.01083
Oxide carbonic	1.00293	Ammonia	1.01881
Marsh gas	1.00634	Sulphurous acid . . .	1.02352
Nitrogen protoxide .	1.00651	Cyanogen	1.02353

All of these gases, therefore, are more compressible than they should be according to the law of Mariotte, and those deviate from it most widely which are readily liquefied.

LAW OF CHARLES.

75. The law for the dilatation of gases, expressed by the formula

$$v = v_0 (1 + \alpha t),$$

is usually called the law of Gay Lussac; but the following criticism by Verdet, which we translate, shows conclusively that it should be called the law of Charles:

"The essential feature of this law, the approximate identity of the rate of dilatation α for all gases, and the consequent proportionality of such dilatations to temperatures indicated by an air thermometer made with any gas, was demonstrated by Charles in a most simple manner. The reservoir of a barometer, filled with the gas, was subjected successively to two temperatures, those of the room and of boiling water; and the rise of the mercury in the tube was observed. For air, oxygen, nitrogen, hydrogen, and carbonic acid, Charles found the ascensions equal; which was all that is necessary to establish the fact that the coefficient of dilatation of these gases is sensibly the same, although its value could not be thus determined with precision.

"To this result, Gay Lussac, who reports the experiments of Charles in his memoir (Ann. de Chim., t. xliii, p. 157), added scarcely anything, except a measurement of the coefficient of dilatation erroneous to nearly $\frac{1}{30}$. It may even be asserted that, in presenting as an absolute law that which is only an approximate relation, he in a measure retarded the progress of science. According to Charles, the *soluble* gases are not dilated as much as air. It is not clear what gases are designated by the term *soluble;* but it is quite probable they may have been those upon which Gay Lussac saw fit to experiment, sulphurous and hydrochloric acid; and for which he announced the coefficient of expansion to be the same as that for air. It is now known to be different to the amount of $\frac{1}{15}$. On this important point, Charles has the advantage; and, however imperfect his method may appear, one that failed to show differences of $\frac{1}{15}$ in the quantity to be measured was not superior."

Though thus criticised, perhaps severely, the experiments of Gay Lussac, which attached his name to the law, have the merit of being the first in which the attempt was made to determine with precision the coefficient of expansion for the different gases. And if he failed, it was not for want of skill or care, but chiefly because of two undiscovered sources of error. At that day, it was not known how a gas and the apparatus containing it can be perfectly dried; moisture, therefore, remained; which, when heated, became vapour and increased the apparent dilatation. Nor was it then suspected that a tube is so imperfectly closed by mercury that air can pass in or out between the glass and the mercury. Gay Lussac used air thermometers, the tubes of which were stopped by short, moveable portions of mercury. After he had measured and obtained for air, oxygen, nitrogen, and hydrogen, between 0° and 100°, the same total dilatation, 0.375, he even imagined that moisture might be a cause of error, and repeated his observations upon air dried by passing it into a thermometer through a chloride of calcium tube; finding again the same number 0.375, he no longer doubted the exactness of his measurements. And, for many years, they were received with universal confidence. Moreover, they were believed to have been confirmed by Dalton, and by Dulong and Petit in 1817; who, although they obtained for 100α the mean value 0.366 between 0° and 300°, adopted without question the value of Gay Lussac 0.375, for temperatures below 100° Centigrade.

The inaccuracy of the experiments of Gay Lussac was first maintained, in the year 1835, by Rudberg, professor in the University of Upsala, in Sweden. Ascribing it altogether to moisture, Rudberg undertook to determine the coefficient with air and apparatus perfectly dried. For this purpose, using air thermometers, as Gay Lussac had done, he, instead of sending the air only once through a chloride of calcium tube, caused it to pass in and out some fifty or sixty times,—expelling it either by heat or by expansion produced by an air pump. He subsequently experimented by a different method, but the only feature peculiar in his researches was the extreme pains taken to render both the air and the apparatus perfectly dry. The mean value given by his results for 100α was 0.3646. To show the importance of getting entirely rid of moisture, Rudberg measured the coefficient for air without drying it, and obtained from one experiment 0.384, and from a second 0.3902; the same apparatus having been again thoroughly dried and then filled with dry air, gave 0.3652 for the coefficient. These results were simultaneously confirmed, in 1841, by Magnus in Berlin, and by Regnault in Paris; the latter of whom then undertook the complete investigation of the dilatation of airs and vapours.

76. We shall give only results and not attempt to describe either the experiments or apparatus of Regnault, referring for such details to the original in the memoirs of the Institute, t. xxi; it is sufficient to say they display extreme accuracy.

If in the thermodynamic equation for gases,

$$pv = p_0 v_0 (1 + \alpha t),$$

the temperature be taken as the independent variable, then the coefficient α may be determined, either directly by observing v under a constant pressure p, or by allowing only p to vary with t; in which latter case the values of v must be calculated by the law of Mariotte. And if that law be not exact, then the values of α obtained by the two methods will differ.

Regnault experimented by five different methods; from the data of four the coefficients of dilatation were deduced from variations of elastic force under nearly constant volume; but in the fifth the variations of volume under constant pressure were directly observed. The following table gives the value of 100α, or expansion from 0° to 100°, for the gases used:

	UNDER CONSTANT VOLUME.	UNDER CONSTANT PRESSURE.
Hydrogen	0.3667	0.3661
Atmospheric air	0.3665	0.3670
Nitrogen	0.3668	0.3670
Carbonic oxide	0.3667	0.3669
" acid	0.3688	0.3710
Protoxide of nitrogen	0.3676	0.3719
Sulphurous acid	0.3845	0.3903
Cyanogen	0.3829	0.3877

This table shows that the coefficient α is greater for the more compressible gases; also that under constant pressure it is greater than under constant volume, except for hydrogen; which is, it will be remembered, less compressible than the law of Mariotte requires.

77. If in the expression for the law of Mariotte,

$$pv = p_0v_0 = p_1v_1,$$

we suppose v to vary as a function of t, while p remains constant, then

$$pdv = p_0 v_0 \alpha \, dt;$$

in which α is the coefficient of expansion under constant pressure.

And if we suppose p_1 to vary with t while v_1 remains constant, then

$$v_1 dp_1 = p_0 v_0 \alpha' \, dt;$$

in which α' is the coefficient of elastic force under constant volume. Hence, by subtraction,

$$pdv - v_1 dp_1 = p_0 v_0 (\alpha - \alpha') \, dt;$$

and integrating,

$$pv - p_1 v_1 = p_0 v_0 (\alpha - \alpha') \, t.$$

If the law of Mariotte were exact, the first member of this equation would be zero; consequently, the coefficients α and α' should be identical. But if the compressibility of the gas increase, or decrease, in a ratio greater than that law requires, then will α be greater or less than α'. We see, therefore, that the law of Mariotte is a *limit* to which permanent gases tend to approach when highly rarefied; and that the reason why the coefficient under constant volume, for hydrogen, exceeds that under constant pressure, is because its resistance or elastic force, like that of a metallic spring, increases with compression, exceeding the ratio of the law of Mariotte; while all other gases act inversely, becoming more compressible the closer their particles are forced to approach each other.

THE COEFFICIENT α VARIES WITH THE PRESSURE.

78. It was announced by Sir H. Davy, in the Philosophical Transactions for 1823, that he had found the coefficient α the same for air at densities varying in the ratios 1, 2, 3, 6, and 12;

and other persons having found that coefficient the same for air at various barometric pressures, it was generally believed to be independent of pressure.

But all such experiments seeming to have been made without sufficient precision, Regnault first examined the dilatation of air of densities both less and much greater than that of the atmospheric pressure, and then extended the investigation to other gases.

For atmospheric air, the following are the results for the coefficient of constant volume:

DENSITY AT 0°.	$1 + 100\alpha'$.	DENSITY AT 0°.	$1 + 100\alpha'$.
0.1444	1.36482	1.0000	1.36650
0.2294	1.36513	2.2084	1.36760
0.3501	1.36542	2.2270	1.36800
0.4930	1.36587	2.8213	1.36894
0.4937	1.36572	4.8100	1.37091

Hence it appears, that between the same limits of temperature, the expansion of air increases with the pressure or density. And for a change of density from 0.1444 to 4.81, that is to say, from 1 to 33.3, the coefficient varied from 0.36482 to 0.37091.

By the methods of constant volume, carbonic acid gave the following results:

DENSITY AT 0°.	$1 + 100\alpha'$	DENSITY AT 0°.	$1 + 100\alpha'$
1.0000	1.36856	2.2976	1.37523
1.1879	1.36943	4.7318	1.38598

The coefficient, therefore, increases much more rapidly with the pressure than it does for atmospheric air.

For the coefficient of constant pressure α, the following values were obtained:

ATMOSPHERIC AIR.		CARBONIC ACID.		HYDROGEN.	
Pressure.	$1 + 100\alpha$.	*Pressure.*	$1 + 100\alpha$.	*Pressure.*	$1 + 100\alpha$.
760^{mm}	1.36706	760^{mm}	1.37099	760^{mm}	1.36613
2525	1.36944	2520	1.38455	2545	1.36616

For hydrogen, the coefficient α does not, therefore, change sensibly between the limits of one and four atmospheres; but it increases rapidly for atmospheric air, and still more so for carbonic acid.

For the small change of pressure from 760^{mm} to 960^{mm}, the coefficient of sulphurous acid, between 0° and 100°, changed from 0.3902 to 0.3980; though the gas at 0°, and under the pressure 960^{mm}, is far from its point of liquefaction.

From the preceding data, Regnault drew the following conclusions: 1°, for the same gas the coefficient of constant pressure is not the same as that calculated from variations of elastic force under constant volume; 2°, the coefficients for different gases are not equal, and the greater the pressure the more marked is this inequality; 3°, air and all other gases, except hydrogen, have coefficients which increase with the pressure; 4°, as the coefficients of expansion of different gases approach nearer to equality when the pressure is diminished, the law of Charles, that all gases have the same coefficient, is a *limit* applicable to gases in a state of extreme dilatation; but which differs the more from reality the denser the gas becomes under compression, or the nearer the molecules approach each other.

INFLUENCE OF TEMPERATURE UPON COMPRESSIBILITY.

79. The results obtained by Regnault for the compressibility of different gases, § 74, were from observations made at temperatures varying from 2° to 10° C.; and are, therefore, absolutely true only under such circumstances.

In order that gases may obey the same law, both of compression and dilatation, for all pressures and temperatures, it is requisite that the ratio of their specific gravities or densities be constant for all such variations. To make this evident, let ω_0 be the weight of any fixed volume of a gas at 0°, and ω'_0 that of an equal volume of atmospheric air at the same temperature, and let both be under the same barometric pressure p_0; let also ω and ω' be their weights for the same volume under the pressure p and at the temperature t. Now, if they both vary by the same law, we shall have for the specific gravity of the gas obtained under these different circumstances,

$$d = \frac{\omega}{\omega'} = \frac{n\omega_0}{n\omega'_0} = \frac{\omega_0}{\omega'_0},$$

a constant quantity, whatever may be the value of n the common factor of variation.

For carbonic acid, Regnault found the following specific gravities or densities, at the temperatures 0° and 100°, and for barometric pressures below that of atmospheric weight:

PRESSURE.	DENSITY AT 0°.	PRESSURE.	DENSITY AT 100°.
760^{mm}	1.52910	760^{mm}	1.52418
374.13	1.52366	383.39	1.52410
224.17	1.52145		

To compare the compressibility of atmospheric air with the law of Mariotte by the same method, Regnault gives the following experimental and calculated weights for the quantity of air contained in a globe at 0°, under different pressures:

PRESSURE.	WEIGHT OF THE AIR IN THE GLOBE AT 0°.		
	Observed.	*Calculated.*	*Difference.*
303^{mm}	5.0895	5.0954	0.0059
312.35	5.2510	5.2522	0.0012
358.22	6.0225	6.0233	0.0008

The differences between the observed and the calculated weights for air are so small that Regnault considered them within the limits of error of observation.

These experiments prove that carbonic acid deviates sensibly from the law of Mariotte at 0° C., even for pressures below that of the atmosphere; but follows that law at 100° C., as air and other permanent gases do, so closely that differences are only observable by methods of extreme precision. It is considered very probable that other liquefiable gases behave as carbonic acid does when compressed at elevated temperatures, but the subject needs experimental investigation.

80. The preceding facts show that gases may be divided into two classes: permanent gases, of which atmospheric air, oxygen, nitrogen, and hydrogen are the most important, and which all obey the laws of Mariotte and Charles so very closely that they may be assumed to do so exactly; and liquefiable gases, whose compressibility increases rapidly with pressure; but at elevated temperatures, or when very highly rarefied, they also appear to follow the law of Mariotte without sensible deviation.

The peculiar exception, which hydrogen presents, of a gas whose compressibility is less than that required by the law of Mariotte, renders it probable that, if the elastic force of other permanent gases be increased by elevating the temperature until the coefficient of elastic force under constant volume becomes equal to that of dilatation under constant pressure, then such gases would follow the laws of Charles and Mariotte exactly; but if the temperature be still further increased, then the difference $(\alpha - \alpha')$ would become negative, as it is for hydrogen, and they too would deviate from that law in the opposite direction. The fundamental law for perfect gases, given by the equation

$$pv - p_0 v_0 (1 + \alpha t) = 0,$$

may, therefore, be properly considered as the expression of a physical state, or *limit*, to which gases approach more or less closely, according to the values of the independent variables p and t, and to that of α, which depends upon the nature of the substance as well as upon the pressure.

HYPOTHETICAL LAW OF RANKINE.

81. As the law of Mariotte does not express exactly the compressibility of any existing gas, it is natural to seek some other formula which would be more correct. In the present state of physical knowledge, it does not appear possible to find one more satisfactory, for the data are very incomplete. Moreover, if we express by the function

$$\phi(p, v, t) = \frac{p_0 v_0}{pv} - 1,$$

the deviation of any gas from the law of Mariotte, which deviation should be zero if that law were exact, then, as the experiments of Regnault show this function to depend upon the temperature, the initial pressure, the variation of the pressure and the nature of the substance, it is improbable that the true function can be of simple form; or that it would be as convenient as the law of Mariotte; which is sufficiently exact for practical uses, and may be considered approximate in nice theoretical investigations.

As the absolute temperature τ is equal to $(a + t)$, and τ_0 is equal to a when t becomes zero, the general formula

$$pv - p_0 v_0 (1 + \alpha t) = 0$$

may be put under the form

$$\frac{pv}{p_0 v_0} = \frac{\tau}{\tau_0}; \tag{99}$$

a remarkably simple form of the equation for the law of Charles and Mariotte, which may be thus enunciated, *the energy of a perfect gas is proportional to the absolute temperature.*

From this expression and "*the hypothesis of molecular vortices,*" Rankine has obtained, for the expansion and elasticity of gases, the formula

$$\frac{pv}{p_0 v_0} = \frac{\tau}{\tau_0} - A - \frac{B}{\tau} - \frac{C}{\tau^2} - etc.;$$

which agrees with the experiments of Regnault, and in which A, B, C are functions of the density to be determined experimentally. (See Rankine on Steam-engines, p. 229, 3d edition.) This equation, regarded as merely an empirical formula, may sometimes be useful; but the hypothesis of molecular vortices is not an accepted part of positive science.

ACCORDANCE OF AIR THERMOMETERS.

82. The variation of the coefficient α, not only for different gases, but also for the same gas under different pressures, suggests the very important question, whether or not temperatures observed with air thermometers made of several gases, or of atmospheric air of variable density, are comparable and exact. This question was fully investigated and settled by Regnault, who arrived experimentally at the following conclusions:

1°, atmospheric air follows the same law of expansion from 0° to 350°, even when its initial elastic force varies from 0.4 to 1.3 metres at 0° ; and consequently, air thermometers are comparable, giving the same indications whatever may be the density of the air with which they are filled ;

2°, atmospheric air, hydrogen and carbonic acid obey, between 0° and 350°, sensibly the same law of expansion, though their coefficients are quite different ; hence thermometers constructed with these several gases accord with each other, provided the temperatures are computed with the coefficient proper to each gas ;

3°, sulphurous acid deviates from the law of dilatation of the preceding gases. Its coefficient of expansion diminishes with the temperature indicated by an air thermometer. Above 100° the sulphurous acid thermometer gives indications which are too small, and the deficiency increases regularly with the temperature. Thermometers made with this gas would, therefore, be incorrect.

The mean coefficient of expansion for sulphurous acid diminishes in a marked manner with the temperature, indicated by the air thermometer. Thus from 0° to 100° it was found to be 0.003825, between 0° and 186° it was 0.003800, and between 0° and 300° it became 0.00379.

From this action of sulphurous acid and the very rapid increase of its coefficient with its density, Regnault was led to infer that most vapours have coefficients very different from that of air when near their points of liquefaction; consequently, under the circumstances in which they are placed for the experimental determination of their densities.

EXPANSION OF VAPOURS.

83. Our knowledge of the expansion of steam and other vapours is exceedingly imperfect. It is customary to apply to them the laws of Charles and Mariotte and the coefficient α of the permanent gases, but such a practice is quite erroneous.

In 1852, Mr. Siemens published experiments on steam, giving the mean coefficient of expansion 0.00693 near the point of ebullition, and a diminishing rate with increase of temperature.

Subsequently, Messrs. Fairbairn and Tate obtained results, which were communicated to the Royal Society and briefly given in the Lond. and Ed. Phil. Mag., vol. xxi, 1861, from which they drew the following conclusions:

1°, that the density of saturated steam at all temperatures, above as well as below 100° C., is invariably greater than that derived from the laws of perfect gases;

2°, that the rate of expansion of superheated steam greatly exceeds that of air for temperatures near the point of saturation; whereas at higher temperatures the rate of expansion approaches that of air and perfect gases. Thus between 80° and 82°, the coefficient for steam saturated at 80° was found to be three times that of air; but at 90° it was nearly the same as that of air.

Messrs. Fairbairn and Tate proposed to extend and complete these experiments. But if this purpose was ever carried into effect, I am not aware that the results have been published.

The expansion of superheated steam has since been investigated experimentally by Hirn, yet without results which can be regarded as final and perfectly satisfactory.

The density of a vapour is independent of its temperature only when its coefficient of expansion is the same as that of air. M. Cahours has drawn the attention of chemists to this fact; which is very important in calculations of the atomic weights of volatile substances from their densities. Acetic acid (crystallized), which boils at 120°, has the following densities:

at 124°, 3.198; at 140°, 2.898; at 190°, 2.378;
" 240°, 2.09 ; " 295°, 2.08 ; " 327°, 2.08.

It is only, therefore, to this final or *limit* density 2.08, that the theoretical formulas apply. Other vapours act similarly.

CHAPTER VI.

INTERNAL ENERGY.

FIRST FUNDAMENTAL LAW.

84. The proposition, that heat is mechanical energy, is very properly called the first fundamental law of thermodynamics; and the equation

$$EQ = \Sigma \int P dp,$$

in which E, the coefficient of Joule, is equal to 425, not only expresses the law, but determines the amount of work equivalent to any given number of thermal units.

We have seen that for hydrogen, oxygen, and atmospheric air, the laws of Mariotte and Charles enable us to calculate E and give for it the value 425, agreeing very closely with that of the experiments of Joule, and confirmed by the working results of engines measured by Hirn. Various experiments made by others all tend to verify the accuracy of this value obtained by Joule.

But it may readily be demonstrated that E is a constant of nature, for which the same value must necessarily be found whatever be the method of determination.

For this purpose, let q be the heat of friction which can be produced by w a given amount of work. If now this heat be reconverted into work, in any engine, it will reproduce the work

$$w = Eq;$$

for if not, then let it produce more or less, expressed by

$$w(1 \pm k).$$

This amount of work will produce by friction a quantity of heat q', determined by

$$w\,(1 \pm k) = Eq'.$$

Now, by proportion, q' converted into work in the same engine gives

$$w\,(1 \pm k)^2\,;$$

and similarly will this give q'', which in the engine will produce

$$w\,(1 \pm k)^3.$$

Thus successively and indefinitely will be obtained q', q'', q''', etc., giving $w\,(1 \pm k)^2$, $w\,(1 \pm k)^3$, $w\,(1 \pm k)^4$, etc.

Hence, if for the supposed variation k the positive sign be taken, the amount of work will increase with each successive transformation of the original amount of energy, or perpetual motion will be realized. But, if the negative sign of k be taken, then energy is annihilated, or unresisted motion is destroyed; which is contrary to the law of inertia and not less improbable than perpetual motion itself.

It is, therefore, evident that the law of Joule is one of those grand laws of nature which are true in all their applications.

SECOND LAW OF JOULE.

85. As thermal action divides itself into internal work, potential and kinetic, U, and sensible or external work, S, we have for it the expression

$$\Sigma \int Pdp = U + S,$$

or

$$dQ = AdU + Apdv.$$

The necessity of eliminating, or determining, the internal work U in thermodynamic investigations is evident from these equations;

and the difficulty of doing so in many cases, see § 54, has already been remarked upon.

Differentiating U in the last equation, as a function of v and t, we have generally

$$dQ = A\left(\frac{dU}{dt}dt + \frac{dU}{dv}dv\right) + Apdv. \qquad (101)$$

Now, for permanent, and consequently for perfect gases, Joule proved experimentally that U is a function of t only; and therefore it does not vary with v; an important law, which shows for such gases that the differential of U as a function of v is zero; and this reduces the last equation to the simpler form

$$dQ = A\frac{dU}{dt}dt + Apdv;$$

by comparing which with the general formula

$$dQ = cdt + ldv,$$

we obtain, for gases obeying this second law of Joule, the consequences,

$$U = Ect; \qquad l = Ap;$$

which may be enunciated thus: *the internal energy of a permanent or perfect gas is a simple function of the temperature; and its latent heat of expansion is directly proportional to the normal pressure.*

Moreover, when there is no external work done by the expanding gas, p is zero, and integrating between limits,

$$Q - Q_0 = c(t - t_0),$$
$$U - U_0 = Ec(t - t_0);$$

equations of fundamental importance, as we have already seen, § 70, for they prove that only air thermometers give indications of temperature proportional to variations of the quantity of heat or internal energy; and they should, therefore, be used as standards for all exact thermometric observations.

As Regnault has shown that c, the specific heat of constant volume, is a constant for each permanent gas, it follows from the second law of Joule, that thermometers made of different gases should indicate the same temperatures, as Regnault has proved experimentally, § 82; for two gases such as hydrogen and air give

$$Q - Q_0 = c\,(t - t_0) = c_1\,(t - t_0);$$

indicating for a given change of heat the same variation of temperature $(t - t_0)$; the only difference being, that linear spaces, graduated into the same number of degrees on the two instruments, vary for the unit of volume in the ratio of c to c_1, or inversely as the specific heats of the two gases.

The second law of Joule, expressed by the formula

$$U - U_0 = Ec\,(t - t_0),$$

is thus enunciated, the internal energy of a gas is a function of the temperature only. It, therefore, does not vary with density or specific volume; and it is of such interest and importance that it will be well to give attention to the experiments by which Joule established this second law. They are also very instructive, for they show clearly the requisites to be attended to in such investigations.

86. If a gas expand without being subjected to external pressure, p is zero and

$$Q = AU;$$

whence by differentiating

$$dQ = A\left(\frac{dU}{dt}\,dt + \frac{dU}{dv}\,dv\right).$$

In this, the first member expresses the quantity of heat requisite to the elementary change of internal state under such circumstances. If now the changes of temperature which take place in the gas be compensatory, or if mere interchanges of heat occur between different portions, then will there be no heat given to, or absorbed from, surrounding bodies; Q and t will be constant and

$$\frac{dU}{dv} = 0; \qquad (102)$$

or the internal energy of the expanding gas is independent of its volume, and is a function of its temperature only.

To subject this theoretical deduction to experimental investigation Joule contrived the following experiment. Two metallic vessels of equal size were connected by a tube provided with a stop-cock; one of them contained air compressed to 22 atmospheres, in the other was as perfect a vacuum as it was possible to produce. When the stop-cock was opened the compressed air from the first vessel flowed into the second, doubling its volume without doing any external work; and the entire apparatus being immersed in water, the quantity of which was reduced as much as possible, it was found that the most delicate thermometer did not indicate any variation of temperature in the enveloping water. The experiment was then varied by immersing each vessel separately; the compressed air, communicating in its expansion sensible motion to its particles, does so at the expense of its internal energy and with consequent reduction of temperature; which is experimentally rendered manifest by immersing it alone; and the air flowing into the second vessel loses its motion, changing kinetic into internal energy with corresponding elevation of temperature. The change of temperature and energy in one vessel is equivalent but opposite to that in the other, and their sum is consequently zero. These experiments of Joule prove that internal energy is independent of volume;

and were the methods sufficiently delicate, they would have left nothing to be desired relatively to the exactness of the law, especially when carefully repeated and verified by an observer so skilled as Regnault; who was led by them to adopt the dynamical theory of heat.

But the quantity of water required to envelope the apparatus was so large, that it may be shown to have been impossible to have detected a change of temperature in the air subjected to experiment to the extent of one degree. It was, therefore, necessary to contrive a much more delicate method of investigation, in order to determine whether the law is rigorously true for any gas, and whether permanent gases obey it alike; or whether, as in the case of the laws of Charles and Mariotte, this second law of Joule is *only a limit* for theoretically perfect gases, to which permanent gases tend to approximate in changes of energy.

87. Such a method was contrived by Sir W. Thomson, and the investigation was then continued by him and Joule jointly. It is evidently necessary to measure the change of temperature of the gas itself, in order to get rid of the concealing influence exerted by the large quantity of water. This was simply done by obstructing the flow of a current of gas through a long pipe, by means of a porous plug; the effect of which obstruction would be to cause on one side of the plug condensation, and on the other side expansion; delicate thermometers show whether the temperature of the gas is changed by the operation. If the gas were perfect, obeying exactly the law of Mariotte,

$$pv - p_0v_0 = 0$$

should be the variation of energy when the compressed gas passes from the pressure p on one side of the plug into the part of the tube communicating with the atmosphere under the diminished pressure p_0 on the other side.

But if the gas does not obey that law, then will

$$p_0v_0 - pv = \delta S;$$

or

$$\delta' Q = A\delta S = A(p_0v_0 - pv)$$

will be the expression for the quantity of heat absorbed in the change of volume due to the escape of the gas into the air. And if the internal work caused by compression and expansion in passing through the porous plug be nothing, then should there be no change of temperature.

But if this be not rigorously true for the gas operated upon,

$$\delta'' Q = A\delta U$$

will express the quantity of heat requisite to this internal change.

Substituting these values in equation (56), we obtain

$$\delta Q = \delta' Q + \delta'' Q = A\delta\,(U + S),$$

for the variation of heat measured in thermal units, which the thermometers should indicate.

The experiments showed for hydrogen a thermal variation which was scarcely appreciable, a very small change for atmospheric air, but a very considerable reduction of temperature for carbonic acid.

They gave the general result, that for each gas the ratio of the reduction of temperature to the difference of pressures is a constant factor,

$$t - t_0 = a\,(p - p_0);$$

hence, if we denote by c the specific heat of the gas,

$$\delta Q = c\,(t - t_0) = ac\,(p - p_0);$$

and

$$\delta U = Eac\,(p - p_0) + pv - p_0 v_0.$$

For an elementary variation of the normal pressure and volume

$$p = p_0 + \delta p, \qquad v = v_0 - \delta v\,;$$

and by equation (98), the formula of Regnault,

$$pv - p_0 v_0 = p_0 v_0\,A'\left(\frac{v_0}{v} - 1\right) + \text{etc.}$$

From these equations, omitting infinitesimals of the second order, we obtain

$$p_0\,A'dv = p_0 v_0\,A'\left(\frac{v_0}{v} - 1\right);$$

but

$$pv - p_0 v_0 = v_0\,\delta p - p_0\,\delta v,$$

and therefore,

$$\delta p = (1 + A')\,\frac{p_0}{v_0}\,dv.$$

Hence, by substitution,

$$\delta U = Eac\,\delta p + p_0\,A'\,\delta v\,;$$

which becomes

$$\delta U = Eac\,(1 + A')\,\frac{p_0}{v_0}\,\delta v + A'p_0\,\delta v.$$

But in this equation the common factor of the second member, $p_0\,\delta v$, is the variation of external work. Dividing by it, we get for the ratio of the change of internal energy to that of external work,

$$\frac{\delta U}{\delta S} = E\,\frac{ac}{v_0}\,(1 + A') + A'\,;$$

an expression in which all the quantities in the second member have been experimentally determined; c by the velocity of sound; A' and v_0 by Regnault; and a and E by Joule and Thomson for atmospheric air, hydrogen, and carbonic acid.

These three gases gave the following values of the ratio of internal to external work:

Atmospheric air	0.0020
Hydrogen	0.0008
Carbonic acid	0.0080

Hence, we see that when air at the usual barometric pressure expands slightly, the internal work is only 0.002 of the external. While for hydrogen it is but two-fifths of that amount, and in fact scarcely appreciable. But for carbonic acid, a liquefiable gas, it is four times greater than for air, and amounts to nearly one per cent.

The effect of elevated temperature was ascertained for air and carbonic acid, and found to be a great diminution of the constant a, or of the rate of cooling to variation of pressure on the different sides of the porous plug. For air, at temperatures of 15° to 20° centigrade, the mean value of the rate a was 0.262; but at 91°.5 it was only 0.206; for carbonic acid at 20° it was 1.151; and at 91°.5 it was reduced to 0.703; comparing these numbers, we see that for carbonic acid the rate reduces from five to three and a half times that of air; thus again manifesting its tendency to approach at high temperatures to the character of a permanent gas.

88. From the above, it is evident that the second law of Joule, like those of Mariotte and Charles, constitutes a limit to which the action of real gases tends to approach when they are highly rarefied; and it is rigorously true only for a theoretically perfect gas.

To comprehend fully its meaning, the analysis of the total action of heat upon an expanding body, already given in § 49, must be borne in mind. That action divides itself into three distinct effects: 1°, change of temperature, rendering the body hotter; 2°, change of internal molecular structure, or variation of the potential of molecular action, which may be called internal work; 3°, external work upon the enveloping surface, or vessel, as that upon a piston pushed by steam, or by expanding air, in the cylinder of an engine. For an elementary thermal change, the last of these three effects is expressed by *pdv;* and the first is denoted by *cdt* in the formula

$$dQ = cdt + ldv,$$

or by the first term of the second member of the equation

$$dQ = A\frac{dU}{dt}dt + A\frac{dU}{dv}dv + Apdv.$$

It is, therefore, only the second term of the last member of this expression, denoting internal work done against cohesion or molecular action, which Joule found to be insensible by his first method of experiment; and which the far superior method of the porous plug proves to be very small, though not inappreciable for air and hydrogen, but quite large for carbonic acid.

Hence it appears that molecular attraction, or the force of cohesion, must be very small in permanent gases, but is quite sensible in a liquefiable gas, such as carbonic acid, and should be nothing for the perfectly gaseous state.

From the equations just used and the second law of Joule, we have already deduced equations (78) for perfect gases,

$$cdt = AdU; \qquad l = Ap;$$

the last being a very simple relation between the mechanical equivalent of heat, the latent heat of expansion and the pressure, from which any one of these three quantities may be determined when the other two are known.

By substituting for l its value Ap, and for dt its equivalent $d\tau$, the variation of absolute temperature, § 71, we have

$$dQ = cd\tau + Apdv. \tag{103}$$

At the absolute zero of heat, both dQ and $cd\tau$ are nothing; and therefore p is so also. Consequently, at that zero, a perfect gas would be without external pressure, without internal motion or temperature, and without molecular attraction or repulsion; or it would be in a state of utter dynamic inaction and indifference, both within itself, and relatively to an enveloping surface.

LAW OF DULONG AND PETIT.

89. The laws of Charles and Mariotte and the second law of Joule being for real gases only limits, or approximations, the question naturally arises, whether the two remaining experimental postulates, from which we have deduced the properties of the perfectly gaseous state, are more rigorously true; or whether they too need to be corrected for perturbations when applied to gases.

To the analyses chiefly of Berzelius we owe the establishment of the great fundamental law of chemistry, that bodies combine in definite proportions by weight; and to Gay Lussac the equally simple law, that gases unite by volumes which are in very simple ratios to each other; water, for example, being composed of *one* volume of oxygen and *two* of hydrogen, and nitric acid consisting of *two* volumes of nitrogen united with *five* of oxygen.

The postulate that *the product of the density by the specific heat of a gas is constant,* see equation (83), is a consequence of the laws of Berzelius and Gay Lussac, coupled with the discovery of Dulong and Petit, that the product of the specific heat of bodies by their chemical combining proportions, or equivalents by weight, is constant for those of like composition. For if gases unite by volume, their combining weights are proportional to their density or specific gravity. Hence, in chemical researches, the combining proportions of gases and vapours are often calculated from their densities.

The usual enunciation of the law of Dulong and Petit is that the product of the *atomic weight* by the specific heat is constant for like chemical substances; but the use of the term *atomic weight* involves the hypothesis of combination by atoms. Whatever may be the probability, or improbability, of that hypothesis, we carefully avoid it, in pursuance of the purpose to exclude from consideration, in discussing the subject of thermodynamics as an exact science, everything purely hypothetical; so that it may carry to your minds the full conviction of necessary truth which only absolute demonstration commands. And this we do the more willingly inasmuch as chemical analyses prove the law of definite combining proportions most positively, thereby rendering the atomic hypothesis of Dalton unnecessary to our purposes, if not even to chemistry itself, and reducing it to the minor importance of being only an ingenious speculation, useful to facilitate by simplifying conceptions of chemical science.

The question of the exactness of the law of Dulong and Petit resolves itself into that of the probable error of measurements of density and specific heat. The requisite manipulations are complex and difficult, tending to increase limits of error. And they are not reliable beyond the third place of decimals. Consequently, we may justly conclude that, in determinations of density and specific heat, the probable error must be such as to render uncertain the third place of decimals.

With these remarks, we submit the following results of Regnault:

	d	c'	$c'd$
Atmospheric air - -	1.0000	0.2375	0.2375
Hydrogen - - - -	0.0693	3.4090	0.2360
Nitrogen - - - - -	0.9713	0.2440	0.2370
Carbonic oxide - -	0.9674	0.2450	0.2370
Carbonic acid - - -	1.5290	0.2164	0.3308
Sulphurous acid - -	2.2470	0.1553	0.3489
Chlorine - - - - -	2.4400	0.1214	0.2962

In this table d denotes the density, c' the specific heat, and $c'd$ their product. It is evident, that while $c'd$ is constant for permanent gases which follow the law of Mariotte closely, such is not the case for vapours and liquefiable gases.

90. We have stated, § 53, that for solids and liquids the capacity, or specific heat, is a function of the temperature. But for permanent gases, Regnault found the specific heat constant at all temperatures and densities. For atmospheric air and carbonic acid, he obtained the following data:

ATMOSPHERIC AIR.		CARBONIC ACID.	
Temperature.	*Capacity.*	*Temperature.*	*Capacity.*
From —30° to +10°	0.23771	From —30° to +10°	0.18427
" 0 to 100°	0.23741	" 10° to 100°	0.20246
" 0 to 200°	0.23751	" 10° to 200°	0.21692

Other gases were experimented upon with like results. Under pressures varying from one to ten atmospheres, Regnault could not find any appreciable change in the capacity of air or of other permanent gases; nor did that of carbonic acid vary with like changes of pressure; though it increases rapidly with the temperature.

In measurements of capacity, or specific heat, that for air 0.2375 is the quantity of heat in calories necessary to raise the temperature of a kilogramme of air, under constant pressure, through one degree centigrade; the specific heat, or capacity, of water being the thermal unit.

91. From these experiments, it follows that the permanent gases are all equally adapted to exact measurements of temperature, comparable with those of thermometers made of atmospheric air; for they all obey closely the law of Mariotte, have nearly the same coefficient of dilatation, and their specific heat, or the quantity requisite to produce a variation of one degree of temperature, is independent of pressure and temperature for each and all of them. But they are the only substances which possess such properties; the liquefiable gases and all solids and liquids have specific heats which increase rapidly with temperature and are not proportional to their expansion. Thus for mercury, according to Regnault, the specific heat is 0.029 between 15° and 20°; but only 0.028 from 10° to 15°; while for alcohol the specific heat is 0.615 from 20° to 15°; 0.602 between 15° and 10°; and 0.596 from 10° to 5°. It is, therefore, only for the permanent gases, that such an equation as (93) can be applicable; or that heat absorbed is not rendered partly latent by the internal work of changing the potential of molecular action. In other words, it is only for them that

$$\frac{dU}{dv} = 0,$$

and that the second member of the equation

$$dQ = A\left(\frac{dU}{dt}dt + \frac{dU}{dv}dv + pdv\right)$$

reduces to the simpler form

$$dQ = cdt + Apdv;$$

in which c is constant for each gas; giving therefore, when there is no external work, the equation

$$Q - Q_0 = c\,(t - t_0),$$

as the law of exact thermometric measurements.

CONCLUSIONS.

92. We have now examined carefully each of the four fundamental postulates, or experimental laws, assumed in § 64, as those of the perfectly gaseous state; they define that state and let us deduce from it simple thermodynamic theorems, which the more permanent gases obey so closely that the differences either fall within the limits of probable errors of observation, or are so small that they may be neglected. Those laws are, therefore, approximations which for practical purposes may be regarded as sufficiently exact expressions of actual phenomena.

In case of hydrogen, we have seen, that while it obeys the last three postulates so closely that it is impossible to detect any difference that may not be considered negligeable, this gas is, with respect to the laws of Charles and Mariotte, (to use an expression of Regnault) *even more than perfect;* for its elastic force increases with pressure, while all other gases become more compressible; and its coefficient of dilatation for constant volume is greater than that for constant pressure; but for other gases it is less. If they, therefore, fall short of the law, hydrogen exceeds it. Hence we may consider, with rigorous exactness, the equation

$$pv = p_0 v_0\,(1 + \alpha t)$$

as the expression of an intermediate state between hydrogen and the other gases, to which they all tend to approach as to a limit.

Moreover, when the liquefiable gases expand with heat, or by removal of pressure, they approximate in their changes to the condition of the permanent gases, and consequently to the perfectly gaseous state.

93. For thermal transformations of gases, we have compared deviations from the laws of Charles and Mariotte to perturbations of theoretically elliptical orbits of planets. With equal fitness have we also compared a perfect gas to the simple pendulum in mechanics; an imaginary thing, for it is impracticable to realize the conditions of its definition, that it vibrate without any resistance in arcs which must be considered cycloidal or infinitesimal; yet, it gives us a simplified equation from which we deduce readily, for vibratory motions, laws which we can generally use as sufficient approximations in practical problems; instead of the more complicated formulas of a real compound pendulum vibrating in a resisting medium.

94. But the justification of our employment of the fiction of a perfect gas as the means of getting simplified approximate laws for real gases, must be put upon higher ground than mere comparison with the successful practice of astronomy, or with examples in rational mechanics. There is not, in fact, any problem in physical mechanics, however simple we may imagine it, which admits of solution in any other manner than by successive approximations; rendered closer by corrections for *residual* phenomena, as Herschel has called them, or facts not taken into account in the first and simpler determination of the principal term or terms of the series. In applying the mechanical theory of heat, or any other, such as that of Fresnel for light, to actual phenomena, we are, therefore, compelled to content ourselves with approximations; and such must necessarily be the case independently of what may be the nature of the phenomena investigated.

If we knew the laws of the internal forces of matter, and those of its molecular structure, it might be possible to give a complete solution of the problem of the effects produced by external forces. Unfortunately, upon this subject we know actually nothing; and if we would avoid errors into which we are led by substituting conjecture for knowledge, we must *eliminate* from our equations, rather than attempt to determine, terms which depend upon internal molecular action, which is so inscrutably concealed from observation.

95. These truths and considerations lie at the very foundation, not only of the study of heat, but also of that division of dynamics into *rational* and *applied* mechanics with which you are familiar. A division of the subject in fact into two sciences,—one purely rational, or abstract, an analytical geometry of *four* dimensions with time and the current co-ordinates of space for variables; and in which the bodies supposed to move are mere fictions, endowed with such properties only as would cause their principal motions to agree with those observed in the material world around us; the other, a science of application, entirely physical, rejecting fiction and hypothesis, absolutely real, positive and practical; the knowledge of the material creation as it actually exists, in so far as it is possible for such knowledge to be discovered and comprehended by man. In every problem of physical mechanics, only partial and imperfect solutions are possible; not of choice, therefore, but of necessity, are planetary motions divided into elliptical revolutions and residual perturbations. Nor is it for mere simplicity and convenience, or in indulgence of fancy, that we begin the investigation of thermodynamic transformations by determining those of a theoretically perfect state, defined as this state has been by the laws of Charles and Mariotte; inductions which the real gases obey quite as closely as do planets move in elliptical orbits, or as machines do ever move without friction.

The full force of these remarks, and the true nature of thermodynamic as well as all other physico-mathematical questions, may be rendered clear by an example, and we select for the purpose the accurate weighing of any body. Nothing seems simpler at first, for we have only to use a good balance and set of weights, and counterpoise the body with such of the weights as will put the whole system into equilibrium. The case is that of a lever with apparently equal arms and equal weights; and for most practical purposes, this first approximation is a sufficient solution of the problem, giving at once

$$mr = wr, \qquad \text{or} \qquad m = w;$$

in which m is the body, w the counterpoise weight, and r, r, the lever arms.

But no heed has been paid to moisture, or other adhering matter. If we attempt to remove this by wiping, the body will be electrified, and may be attracted by other adjacent substances.

Then the apparent weight must be corrected for atmospheric buoyancy; w is not the true weight of m; and we must add a term β for buoyancy, giving

$$m = w + \beta + \text{etc.};$$

in which β is to be ascertained by obtaining the specific gravity of air under the particular barometric pressure, temperature and degree of moisture, at the time and place of weighing m, and the difference of volume of m and w at the same temperature. Denoting the specific gravity of the air by a, we have

$$m = w + a(v - v') + \text{etc.}$$

To ascertain the values of a and of v and v', we must solve problems equally if not more difficult than that of weighing m, and requiring corrections. Thus rapidly does the question grow complicated.

But suppose the correction for buoyancy made, either approximately or exactly, if to do so exactly were possible; how are we to know that the weight w is not false, otherwise than by comparing it exactly with a true standard, or know this standard to be true, except by proving it so by complex investigations? Assuming this error of w ascertained, we must correct both w and β for it, giving

$$m = w + \beta + \delta(w + \beta) + \text{etc.}$$

Thus far the balance has been supposed perfect, and its lever-arms exactly equal, conditions which are neither of them possible. All the imperfections of the balance must, therefore, be determined, as well as the difference of its lever-arms at the time of weighing m; and for these corrections must be applied. So, by successive approximations, we arrive more and more closely at the probable weight of the body, determining the principal terms which are to be added in the second member of our expression.

Finally, after having made all possible corrections, we reach the question of errors of observation, errors of dynamic change of temperature, volume and moisture, during the time of the experimental investigation, instrumental errors that cannot be ascertained, etc., etc., and must then content ourselves with simply determining the *limit* of probable error, or degree of the approximation in the obtained result.

CHAPTER VII.

AIR ENGINES.

ELIMINATION OF INTERNAL ENERGY.

96. We have already remarked that, in thermodynamic investigations, our ignorance of the internal energy and nature of bodies presents a difficulty so great that, unless there were some mode of eliminating it, it would be quite insurmountable.

Sadi Carnot first showed that, to overcome this difficulty, it is necessary that the body, or system, return by any series of changes back to its primitive state.

The first fundamental law gives, for a body expanding by heat between the states (1) and (2), the equation

$$E(Q_2 - Q_1) = U_2 - U_1 + \int_1^2 pdv.$$

But should the body pass by any cycle, from the state (1) back to the same initial state (1), then will

$$E(Q_2 - Q_1) = \int_1^2 pdv \, ; \qquad (104)$$

or the variation of heat will be entirely converted into external work; the internal work being completely eliminated, or compensated, for

$$U_2 - U_1 = U_1 - U_1 = 0.$$

And this is evidently true for all substances, as well as for all cycles.

Such, therefore, is the simple condition, by the fulfilment

of which internal work or energy may be eliminated and thermodynamic transformations become measurable by external work alone.

In steam-engines, the operations consist of a perpetually recurring cycle of precisely similar positions, states and temperatures; they, therefore, evidently fulfil the condition of Carnot, and equation (104) is directly applicable to them.

WATT'S INDICATOR AND CLAPEYRON'S DIAGRAM OF ENERGY.

97. Of the many valuable contrivances relating to the steam-engine, for which the world is indebted to the inventive genius of James Watt, one of the most simple and beautiful is his indicator; a little instrument, by which an engine is made to furnish a drawing of its own work, or graphically to integrate equation (104). This indicator, attached usually to the working cylinder, consists of a small cylinder and piston, to the rod of which is fastened a pencil, moveable by the pressure of the steam but resisted by a spiral spring; so that it records upon a paper touching the pencil-point changes of pressure as variation of the ordinates of a curve. Another motion is given to the paper perpendicularly to that of the pencil, which causes it to mark changes of volume, or variations of the abscissas of the curve. This latter motion is so connected with that of the engine as to be made reciprocating, causing the paper to reverse its direction periodically, so that the diagram becomes a closed curve or cycle; the area of which represents the integral of pdv, or the external work.

For a more complete description of this little instrument, reference may be made to Pambour, "Théorie des machines à vapeur," page 109, or to some other treatise on the steam-engine.

We owe to Clapeyron this use of Watt's indicator and diagram of energy, as the graphic method of applying the principles of Carnot to the work of engines.

CARNOT'S TEST OF PERFECTION.

98. An engine always receives from a boiler, or source, at a temperature τ, a quantity of heat which we may denote by q; and it emits at a *lower* temperature τ_0 a part q' to a condenser, or refrigerator, which in non-condensing engines is the atmosphere. If, of the difference $(q - q')$ none be lost or wasted, but all be converted into useful work, then will the engine evidently be theoretically perfect. This, according to Carnot, would be the complete utilization of the *chute de chaleur* from τ to τ_0, or from the higher to the lower temperature. Carnot, however, was led by the material hypothesis to suppose erroneously q equal to q'; for according to that hypothesis heat cannot be put out of existence; while by the dynamical theory it ceases to be heat, or to exist as thermal energy, when transformed into mechanical work.

Carnot's test of a perfect engine is that it be capable of being worked backwards, in the cycle, reproducing all the changes; or *that it be reversible.* An expanding gas converts internal heat, or energy, into external work, and when compressed by external pressure grows hot; as these opposite changes are conversely equivalent, it is evident that if an engine worked by heat produces the amount of work

$$E(q - q') = \int_1^2 pdv,$$

so would it, if worked backwards, by expending this mechanical work upon it, produce the amount of heat

$$q - q' = A \int_1^2 pdv;$$

provided that each of these transformations take place without waste or loss. It is also necessary that the changes, or phenomena, be always such as are capable of being reversed; but in some cases this is impracticable. For example, bars of iron are heated by friction or by hammering them; but hot iron cannot be made to perform directly any such work as rubbing or hammering.

99. To prove that reversibility is the test of perfection in engines, let it be supposed that some other engine receiving heat from a hot body or source A and giving it off to a refrigerator B, can, for the same temperatures τ and τ_0, do more work than a reversible one. Then the two may be combined into one compound engine; the first receiving and conveying a certain quantity of heat, works the reversible engine backwards, causing it to take from the refrigerator the same heat and convey it to the source or hot body A; thus producing a perpetual performance of work without expenditure of energy, or in other words perpetual motion would be realized.

It is evident, therefore, that to perform work the reversible engine would have to convey from the refrigerator a quantity of heat *greater* than is conveyed to it by the first engine; or that work must be performed by a cold body giving heat to a hot one, or by cooling itself below the temperature of surrounding bodies; a mode of producing work which is contrary to all knowledge of the phenomena of heat and clearly impossible.

CYCLES OF CARNOT AND CLAPEYRON.

100. Among the various cycles of thermodynamic transformation, or diagrams of energy, there is one especially important in the discussion of the work of engines, and which is usually called the cycle of Carnot.

In this cycle, the work is represented by an area bounded by four intersecting curves; two *isothermal* lines, or curves of constant temperature, cut by two *adiabatic* lines, or curves of constant quantity of heat, sometimes called lines of *no transmission.* The term *adiabatic,* proposed by Rankine, seems preferable.

If in the annexed diagram of energy, we suppose a hot body to pass from the state or condition of temperature, pressure and specific volume, denoted by the position *M,* whose co-ordinates are p and v, the corresponding pressure and specific volume, to the state of the position *N,* passing through all the intermediate states of the *isothermal* line *MN,* for which the absolute temperature τ is constant, and

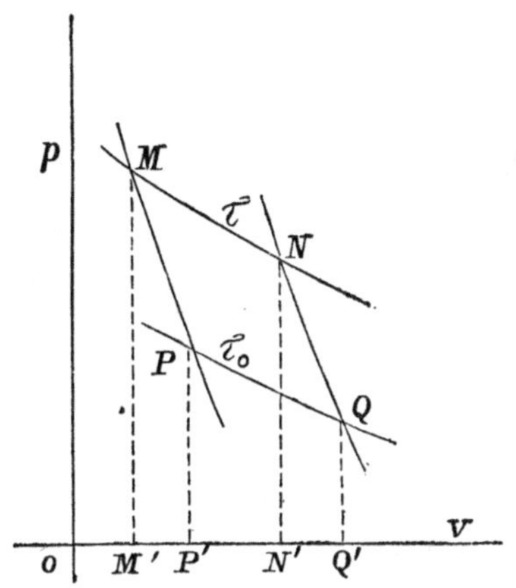

$$\tau = f(p, v) = c;$$

then that it pass from the state *N* to that of *Q* by the successive changes indicated by the *adiabatic* line *NQ;* it will perform during these dynamic changes an amount of *positive* work of expansion determined by integrating the expression pdv, first between the limits of *M* and *N,* and next between those of *N* and *Q.*

If now we suppose the engine to perform upon the substance an amount of *negative* work of compression; first from *Q* to *P* through the states of constant temperature denoted by the isothermal line *PQ,* whose equation is

$$\tau_0 = f_0(p, v) = c_0;$$

then by the *adiabatic* line *PM,* for which the body neither receives nor gives heat to surrounding substances, while its pressure, volume and temperature vary with the work performed upon it; then the body will thus have returned to the primitive state M by the cycle $MNQP$, a cycle of Carnot.

The external work for this cycle will be the integral of pdv taken for all points of the area $MNQP$, positively from M to Q, negatively from Q back to M; moreover, it will be equal to

$$E(q - q') = \int_0^0 pdv, \tag{105}$$

or the difference of the quantities of heat received from the source at the temperature τ and given at the lower temperature τ_0 to the refrigerator, multiplied by E, the equivalent or coefficient of Joule.

It will also be a *maximum*, for in the *chute de chaleur* from τ to τ_0 there is no loss of heat during the transformations of the adiabatic lines NQ and MP, and as the body returns to its initial or primitive state M the internal work is zero; or the *chute de chaleur* is wholly converted into external work.

To realize such a cycle of transformations, it is necessary: 1°, that the substance acted upon, water or steam in ordinary engines, be in contact with a perfectly conducting hot body, the boiler or source, freely receiving from it the quantity of heat q, at the constant temperature τ, along the isothermal line MN; 2°, that, while it expands from N to Q and works along the adiabatic line NQ, it be surrounded by perfect non-conductors of heat; 3°, that, between Q and P, it be in contact, at the constant temperature τ_0, with the refrigerator, or condenser, which shall freely abstract from it the quantity of heat q'; 4°, and finally, that along the adiabatic line PM it be again completely enveloped by perfect non-conductors.

Such are the conditions requisite to constitute a perfect engine, capable of utilizing, or converting entirely into external work, the difference $q - q'$, or amount of thermal energy consumed by it. But as there are no substances which are perfect non-conductors of heat, the conditions required by adiabatic curves are practically impossible. All engines are, therefore, rendered more or less imperfect by radiation and conduction; the more completely, however, the pipes, cylinder, etc., can be rendered non-conducting, the greater will be the economy.

101. As the total quantity of heat received from the source is q, and of this q' must be given up to the refrigerator without recovery, it is evident that *the duty*, as it is technically called by engineers, or the maximum efficiency of a perfect engine is to be measured, not by the quantity of heat communicated q, or by the gross amount of fuel consumed, but by

$$e = \frac{q - q'}{q}, \tag{106}$$

the ratio of the heat converted into work to the total quantity received.

102. The reversibility of a cycle of Carnot may be readily shown. Let the direction of the cycle of operations be reversed, then from M to P the substance will expand without receiving heat, but performing the amount of positive work graphically represented by the area $MM'PP'$; next it will pass at the temperature τ_0 by the isothermal line PQ to Q, receiving the quantity of heat q', and converting it into work $PP'QQ'$; then from Q to N, without change of quantity of heat, it is compressed by *negative* internal work done upon it, graphically denoted by $NN'QQ'$, with increase of temperature or internal energy from τ_0 to τ_0; and lastly, by the negative work $NN'MM'$ along the *isothermal* line NM, it is compressed back to M and gives off at τ the heat q. The total change of work will be again indicated by the difference of positive and negative areas; but will now be *negative* and equal to $MNPQ$; and the total heat *emitted* will be $q - q'$; so that more work is expended upon the engine than it performs, and this excess of work is transformed into the heat

$$q - q' = A\int_0^0 pdv; \qquad (107)$$

which is identical with (105), the equation for the changes in the opposite direction. In the first case, therefore, when operating directly the engine transforms heat into mechanical work; in the second, work is converted into heat; and in both the relative quantities are the same, or they are equivalent.

AIR ENGINES.

103. As practical examples of engines working in simple cycles, we may take the hot-air engines of Stirling and Ericsson; in which the ingenious artifice of employing a *regenerator*, or *economizer*, was resorted to, in order to overcome the serious imperfection of loss by radiation and conduction, or the impossibility of an adiabatic line. The office of the regenerator is that of restoring to the body transformed, during compression from τ_0 to τ, the amount of heat lost by couduction of enveloping surfaces during expansion from τ to τ_0. Thus seeking to prevent loss by conduction and radiation, and to render the engine perfect, with a *chute de chaleur* from τ to τ_0; which fall of temperature may evidently be made much greater for a hot-air engine than for a steam engine; as the temperature τ may be that of incandescence, or redness, without danger of explosion.

AIR ENGINE OF STIRLING.

104. This engine was invented in 1816, and is very ingenious. For full and satisfactory descriptions of it, we must refer to works illustrated by well executed working drawings, as it forms no

part of the purpose of this elementary course to give plans for the construction of engines. We limit ourselves to the simple exposition of the laws of heat, and give for their application only such diagrams as are necessary to render the subject clear and intelligible.

In the engine of Stirling, the body or mass of air is first heated, under constant volume, from the state of pressure and temperature at A to that of B; increasing its temperature and pressure until it becomes able at the state B to overcome the resistance of the piston. From B to C the air expands under constant temperature τ, following the isothermal line BC. From C to D its temperature is lowered from τ to τ_0, while its volume remains constant and the pressure decreases. Lastly, under the constant temperature τ_0, while acted upon by the refrigerator and compressed by the engine, it passes from D back to the original state at A. During the changes from A to B and from C to D, there is no variation of volume and therefore no external work; but between A and B the temperature is elevated, requiring for that effect the amount of heat

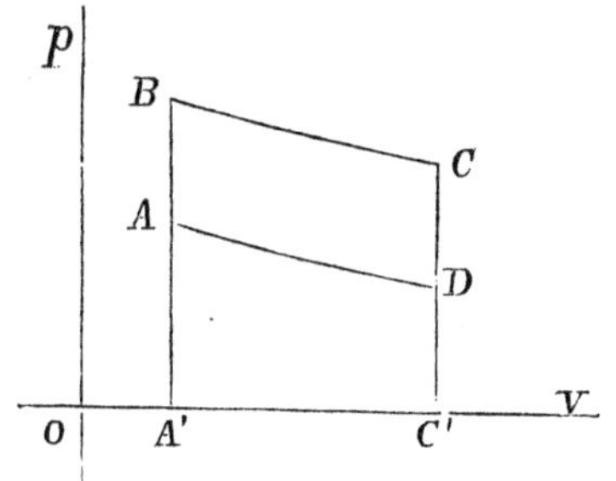

$$c(\tau - \tau_0),$$

in which c is the specific heat of air under constant volume. Again from C to D there is cooling, or liberation of heat, to an equal extent, giving out

$$c(\tau - \tau_0);$$

and it is the office of the *regenerator*, or *economizer*, to render the heat emitted from C to D available for the change from

A to B; so that, in the integration of equation (103), which is applicable to this engine, or of

$$dQ = cd\tau + Apdv,$$

the definite integral of $cd\tau$ shall reduce to zero by compensation of equal loss and gain.

The formula for the engine thus becomes

$$dQ = Apdv;$$

which is readily integrable, for the isothermal line BC is an hyperbola referred to rectangular asymptotes, and its equation is

$$pv = p_1v_0 = p_0v_0\,(1 + \alpha t). \qquad (a)$$

Similarly, the equation of the isothermal line AD is

$$pv = p_0v_0\,(1 + \alpha t_0). \qquad (b)$$

Hence, the amount of positive external work done by the change from B to C will be represented graphically by the area $A'BCC'$, requiring the amount of heat q to be absorbed from the source or fire between B and C. And algebraically, we have

$$q = A\int_{v_0}^{v_1} pdv;$$

in which v_0 is the volume indicated by the abscissa OA', and v_1 that of the abscissa OC'. Along the isothermal line DA, a quantity of heat q' is given out to the refrigerator, the work is negative, represented by the area $ADC'A'$ and algebraically

$$q' = A\int_{v_0}^{v_1} pdv.$$

The total work of the engine will be represented by the quadrilateral area $ABCD$ and may also be expressed by

$$E(q - q').$$

To determine its value, equations (*a*) and (*b*) give

$$q = Ap_1v_0\int_{v_0}^{v_1}\frac{dv}{v} = Ap_1v_0 \log\frac{v_1}{v_0}.$$

$$q' = Ap_0v_0\int_{v_0}^{v_1}\frac{dv}{v} = Ap_0v_0 \log\frac{v_1}{v_0}.$$

Consequently, we have for the duty, or maximum efficiency, of the engine the ratio of areas

$$\frac{ABCD}{A'BCC'} = e,$$

or the algebraic formula

$$\frac{q - q'}{q} = \frac{p_1 - p_0}{p_1} = \frac{\tau - \tau_0}{\tau}. \qquad (108)$$

For, since p_1 and p_0 are the pressures under the same volume v_0 of the same mass of air at the temperatures τ and τ_0, the law of Charles and Mariotte gives

$$\frac{p_1v_0}{p_0v_0} = \frac{\alpha(a + t)}{\alpha(a + t_0)} = \frac{\tau}{\tau_0}.$$

The expression for the duty, or maximum efficiency, of an engine, working without loss,

$$\frac{q - q'}{q} = \frac{\tau - \tau_0}{\tau} = e \qquad (109)$$

gives the ratio of the heat used to that received, or of the equivalent *chute de chaleur* $(\tau - \tau_0)$ to τ; and we shall find this ratio, or theorem, to be one of the greatest importance, applicable to all heat engines, and in fact constituting the second fundamental law of thermodynamics.

ERICSSON'S ENGINE.

105. This engine may be theoretically considered a very interesting modification of that of Stirling. In its cycle the air is heated under constant pressure, instead of being heated under constant volume.

In its diagram of energy, the body passes from the state A to to that of B, under the constant pressure OM, and with the change of volume AB. The quantity of heat required for the transformation is

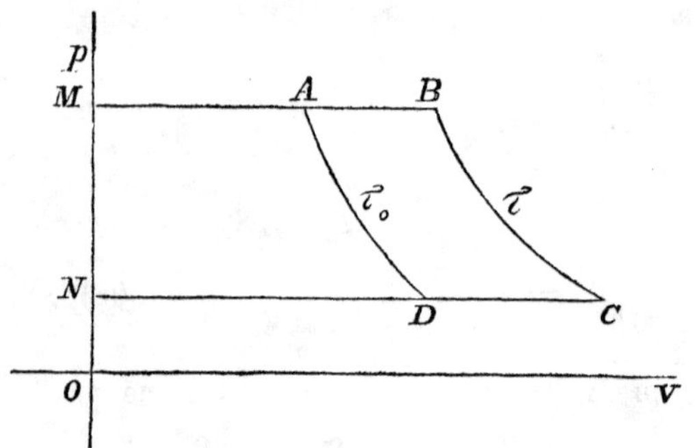

$$c' (\tau - \tau_0);$$

in which c' is the specific heat of air under constant pressure. From B it then passes to the state C under the constant temperature τ, and by the isothermal line BC, absorbing in the change the quantity of heat q; from C it is compressed by the engine to the state D, and passing from the higher temperature τ to the lower τ_0 it emits the quantity of heat

$$c' (\tau - \tau_0).$$

From D it is compressed back again to A; and being at the same time in contact with the refrigerator, it gives to it the quantity of heat q'. The heat utilized will be $(q - q')$; and the regenerator is employed to produce the compensation of loss and gain of the quantities of heat, each equal to

$$c' (\tau - \tau_0),$$

in the changes from A to B, and from C to D.

It is evident from the diagram, that the work done will be denoted by the area $ABCD$, and the *duty* by the ratio of that area to $MNCB$. Algebraically, the quantities of heat received and given out are

$$q + c'(\tau - \tau_0),$$
$$q' + c'(\tau - \tau_0).$$

The hyperbolic areas are

$$p_0 v_1 \log \frac{p_0}{p_1},$$

$$p_0 v_0 \log \frac{p_0}{p_1},$$

and the area $ABCD$ has for its value

$$p_0 (v_1 - v_0) \log \frac{p_0}{p_1}.$$

The *duty*, or maximum efficiency, is

$$\frac{q - q'}{q} = \frac{v_1 - v_0}{v_1} = \frac{\tau - \tau_0}{\tau}.$$

Consequently, the efficiency would be the *maximum* for both of these engines if no heat were uselessly wasted in changes of temperature; the quantity $c(\tau - \tau_0)$ in one, and $c'(\tau - \tau_0)$ in the other, being economized and made to circulate in successive cycles, by being alternately given to and taken from the regenerator or economizer.

106. Thus far we have considered engines supposed to be perfect, and have sought to indicate the conditions necessary to render them so. It is evidently important to determine such conditions, and that an engineer be familiar with them, for they enable him to detect imperfections and suggest improvements.

An engine is perfect if it fulfil Carnot's criterion of reversibility, or have for its coefficient

$$\frac{q - q_0}{q} = e = \frac{\tau - \tau_0}{\tau};$$

which requires that it receive the quantity of heat q at the temperature τ only, and part with q° to the refrigerator only at τ_0, neither receiving nor emitting heat at any other temperatures.

But as such an engine, working thus by a cycle of Carnot, is an impossibility, it becomes important to study by what contrivances engines may be made to economize the waste of power due to absorption and emission of heat at other temperatures than τ and τ_0, or from other bodies than the source and the refrigerator.

Of such contrivances, none is more ingenious than the regenerator of Stirling. If this regenerator rendered the compensation complete, evidently the engine would be as perfect as one working in a cycle of Carnot.

The corresponding diagram of energy would be a quadrilateral area, bounded by two isothermal lines intersected by two lines of *equal* loss and gain, instead of two *adiabatic* lines; to such lines of *equal* transmission Rankine gives the name *isodiabatic.*

ISODIABATIC LINES.

107. To express symbolically the relation of lines which are isodiabatic. Let q be the heat absorbed and q' that emitted during the changes between τ and τ_0, or for the isothermal lines MP and NQ. Now for any infinitesimal corresponding changes along the isodiabatic lines MN and PQ, we have

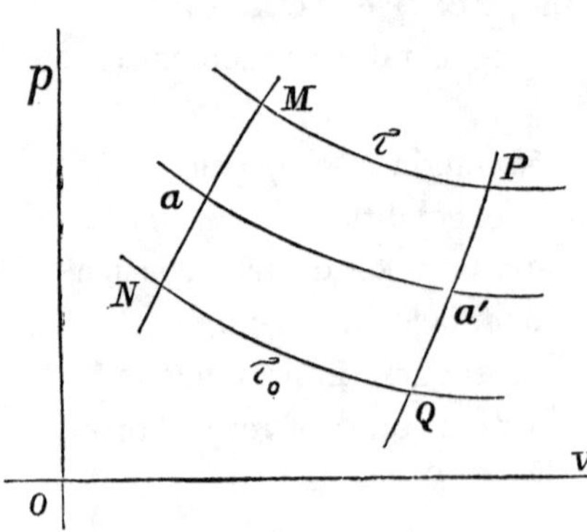

$$dq = cd\tau + Apdv,$$
$$dq' = cd\tau + Ap'dv';$$

and if these be equal, then must

$$pdv = p'dv'.$$

These corresponding changes taking place from positions a and a' on an isothermal line, we may assume for permanent gases the law of Mariotte to hold good, therefore

$$pv = p'v',$$

and

$$\frac{p'}{p} = \frac{v}{v'} = \frac{dv}{dv'} = n.$$

If, therefore, for any line of change MN, we know its equation

$$f(pv) = 0,$$

we have only to substitute in this the values

$$v = nv', \qquad p' = np,$$

to obtain for the isodiabatic line PQ its equation

$$f\left(\frac{p'}{n}nv'\right) = 0.$$

In the engine of Stirling, the volume during the change from N to M is constant, and its equation is therefore

$$v - v_0 = 0;$$

consequently, the equation of PQ is

$$nv' - v_0 = 0;$$

and both are evidently straight lines parallel to the axis of p.

In Ericsson's engine, the pressure is constant for MN, and its equation is

$$p - p_0 = 0;$$

hence for the equation of PQ, we have

$$p' - np_0 = 0;$$

and both are straight lines parallel to the axis of v.

It is evident that the general equation of the line MN is arbitrary; and consequently, that the problem may have an infinite number of solutions.

Theoretically, a regenerator should absorb heat only from the hot air and give it back to it afterwards without loss; practically, it is impossible to prevent waste; a large portion of the heat being always communicated by conduction and radiation to surrounding bodies. The term *economizer* is, therefore, its appropriate name. And we see that for air and all other heat engines, loss by conduction and radiation must ever render it impracticable to make them dynamically perfect. Yet of such loss a regenerator may save quite a large part.

One of the greatest improvements made in furnaces for metallurgy is that of Mr. Siemens, who, by attaching a *regenerator* composed of a mass of fire-brick, has introduced great economy of fuel united with other very important advantages.

ADIABATIC CURVES.

108. To proceed further, it is necessary that we investigate the nature and properties of those lines of no transmission called by Rankine *adiabatic;* and which with *isothermal* lines form the diagrams of energy in cycles of Carnot.

For solids and liquids, our ignorance of the functions which express the relations of pressure, volume, and temperature during thermodynamic changes, renders it quite impossible to determine the equation of an adiabatic curve. But for perfect, and consequently for permanent, gases the equation is easily obtained.

For perfect gases, equation (80) gives the relation of the specific heats of constant pressure and constant volume,

$$c' - c = A\,\alpha\, p_0 v_0 \cdot$$

and if v and t be taken as independent variables, equation (88) gives

$$\frac{dQ}{a+t} = \frac{cdt}{a+t} + A\,\alpha\, p_0 v_0 \frac{dv}{v}.$$

Denoting the first member of this by $d\phi$, substituting in the last term of the second member, and integrating, we find

$$\phi = \log \beta\,(a+t)^c\, v^{c'-c}.$$

And replacing $(a+t)$ by its value given by

$$pv = \alpha\, p_0 v_0\,(a+t),$$

we get

$$\phi = \log \frac{\beta}{(\alpha\, p_0 v_0)^c}\, p^c v^{c'}.$$

But as the constant of integration β is arbitrary, we may put

$$\beta = (\alpha\, p_0 v_0)^c,$$

and our equation then becomes

$$\phi = \log p^c v^{c'}. \qquad (110)$$

By comparing this result with equations (75) and (89), we see that

$$d\phi = \frac{dQ}{\lambda} = \frac{dQ}{\tau},$$

and

$$d\phi = d \log p^c v^{c'}. \qquad (111)$$

109. When the specific volume, or density, of a gas changes without heat being either absorbed or emitted, dQ is zero, and the product $p^c v^{c'}$ is constant; this is the law of Poisson, and it may be expressed thus,

$$p^c v^{c'} = p_0^c v_0^{c'}; \tag{112}$$

the following modification of which is often used

$$pv^\gamma = p_0 v_0^\gamma; \tag{113}$$

in which

$$\gamma = \frac{c'}{c},$$

the ratio of the two specific heats or capacities.

The equations just found replace the law of Mariotte, whenever a perfect, or a permanent, gas varies in volume, pressure, and temperature without receiving, or imparting, heat to other bodies. And they are evidently those of a line of no transmission, or of an adiabatic curve.

The form of its equation shows an adiabatic curve for airs to be a hyperbola referred to asymptotic axes; but for which the ordinates p vary more rapidly than the abscissas v; because the specific heat under constant pressure c' is always greater than c, the specific heat of constant volume.

The law (113) was demonstrated by Laplace and Poisson before the dynamical theory of heat was accepted; and it is indeed independent of any ideas we may conceive of the nature of heat.

When heated air, or superheated steam, expands in the cylinder of an engine, after being *cut off*, it may be approximately considered as changing by the law of an adiabatic curve, if it varies so rapidly that time is not allowed for loss of heat by conduction.

For other substances than perfect or permanent gases, it has not been found possible to determine the form of the function ϕ; but equation (75) and the condition dQ equal to zero give as a general expression of no transmission of heat, or of an adiabatic curve,

$$\phi = \int \frac{dQ}{\lambda} = \text{a constant}; \tag{114}$$

which is, therefore, true for all substances; though the function ϕ is not the same constant for different bodies, but depends for each upon its particular nature.

CURVES OF TRANSFORMATION.

110. Three kinds of lines, or curves of thermodynamic change, are employed: 1°, adiabatic curves; 2°, isothermal lines; 3°, lines of equal energy. For perfect gases these lines reduce to two kinds only; for the isothermal lines are those of equal energy. This is evident from the fact that, by the second law of Joule, the internal energy of a gas is a function of its temperature alone; giving

$$cdt = AdU, \qquad \text{or} \qquad U = f(t).$$

Hence, in every change in which the temperature of a gas remains constant, the internal energy does not vary; and isothermal lines for a perfect gas are, therefore, lines of equal energy.

For the perfectly gaseous state, we have, for an isothermal line, the law of Charles and Mariotte, or the equation

$$pv = p_0 v_0 (1 + \alpha t);$$

in which the temperature t is constant for the same line. And for bodies generally, we have for an isothermal line

$$t = f(vp) = c,$$
$$Mdv + Ndp = 0.$$

For adiabatic lines, perfect and permanent gases give the law of Poisson,

$$p^c v^{c'} = p_0{}^c v_0{}^{c'};$$

and generally for all substances,

$$\phi = \int \frac{dQ}{\lambda} = \text{a constant.}$$

If this integral be taken between any two limits or states (1) and (2), its value is

$$\int_1^2 \frac{dQ}{\lambda} = 0. \qquad (115)$$

The geometric construction of an isothermal line is for perfect gases an equilateral hyperbola, with p and v for asymptotic co-ordinates. That of an adiabatic line for such gases is also an hyperbola referred to asymptotic axes; but as p varies more rapidly than v, the curve recedes more from the axis of v than it approaches that of p, and is not, therefore, symmetrical either in position or form. For liquids, their very slight compressibility shows that the ordinates p vary far more rapidly than the volumes v; and for solids, the outward pressure, or elastic repulsion, changes even to attraction, so that p becomes negative, and the curve cuts the axis of abscissas.

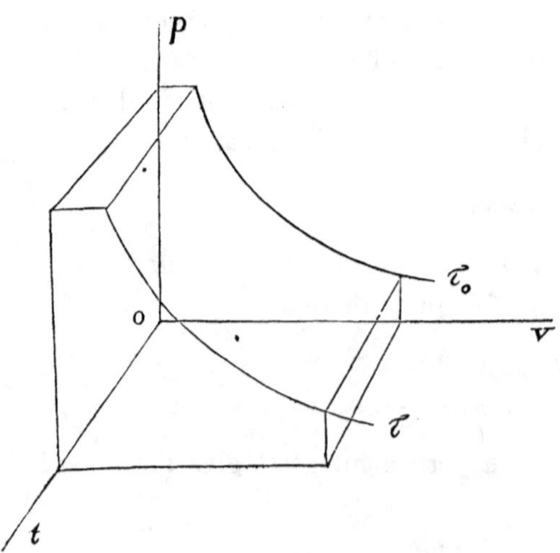

The law of Charles and Mariotte may be put under the form

$$pv = R\tau,$$

which is the same as

$$xy = cz.$$

Its geometric construction is the hyperbolic paraboloid, represented in the annexed figure.

When τ is constant the corresponding section gives for the isothermal line

$$pv = xy = a,$$

an equilateral hyperbola. And for sections perpendicular to the axes of p and v respectively, we have

$$v = x = a't,$$

$$p = y = a''t,$$

or the intersections of the planes with the paraboloid are straight lines.

AIR ENGINES WORKING IN CYCLES OF CARNOT.

111. In the accompanying diagram let the quadrilateral $MNPQ$ represent a cycle of Carnot. The isothermal lines of the temperatures τ and τ_0 being MN and PQ; and MP and NQ, or ϕ_0 and ϕ, being adiabatic lines.

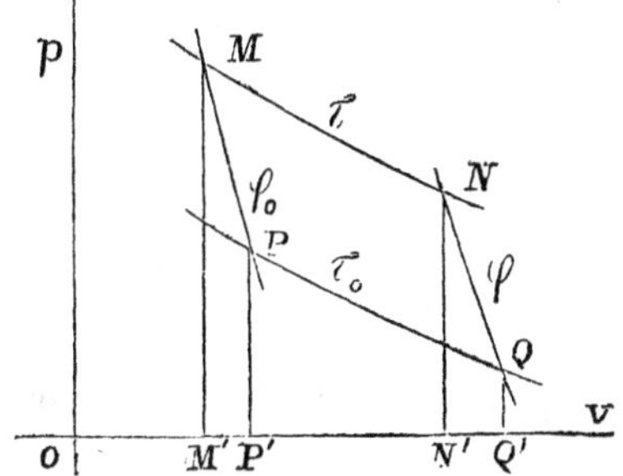

In the change from M to N, at the temperature τ, the quantity of heat q is absorbed; and q' is the quantity given at the temperature τ_0 to the refrigerator in the change from Q to P. Between N and Q, and again from P to M, no heat is either emitted or received, the curves ϕ and ϕ_0 being adiabatic.

The integral of *positive* work is graphically represented by the sum of the two areas $M'MNN'$, and $N'NQQ'$. The work from N to Q is done by expansion after communication is cut off at N from the source of heat.

The *negative* work, or that of compression, done by the engine upon the air from Q to P and from P back to M, is represented by the areas $P'PQQ'$ and $M'MPP'$.

The efficiency is measured by

$$\frac{q-q'}{q}=\frac{MNQP}{M'MNN'};$$

and it is required to prove that this ratio is equal to

$$\frac{\tau-\tau_0}{\tau}.$$

Let p_0, p_1, p_2, p_3 be the respective pressures at P, M, N and Q.

For the heat received from the source, between M and N, we have

$$q=Ap_1v_1\log\frac{p_1}{p_2};$$

and for the heat given to the refrigerator, during the change of state from Q to P,

$$q'=Ap_0v_0\log\frac{p_0}{p_3};$$

From N to Q the equation (112) of an adiabatic line gives

$$v_2p_2^{\frac{c}{c'}}=v_3p_3^{\frac{c}{c'}}; \qquad (a)$$

but by the law of Charles and Mariotte,

$$p_2v_2=ap_0v_0\tau,$$

$$p_3v_3=ap_0v_0\tau_0;$$

and therefore

$$\tau_0v_2p_2=\tau v_3p_3.$$

Dividing this equation by (a) we get

$$\tau_0p_2^{\frac{c'-c}{c'}}=\tau p_3^{\frac{c'-c}{c'}};$$

which may be put under the form

$$\frac{p_2}{p_3} = \left(\frac{\tau}{\tau_0}\right)^{\frac{c'}{c'-c}}$$

In like manner, we obtain, between P and M,

$$\frac{p_1}{p_0} = \left(\frac{\tau}{\tau_0}\right)^{\frac{c'}{c'-c}};$$

consequently

$$\frac{p_1}{p_2} = \frac{p_0}{p_3};$$

and the efficiency is

$$\frac{q - q'}{q} = \frac{p_1v_1 - p_0v_0}{p_1v_1};$$

but

$$p_1v_1 = \alpha\, p_0v_0\tau,$$

$$p_0v_0 = \alpha\, p_0v_0\tau_0;$$

and these values give, by substitution,

$$\frac{q - q'}{q} = \frac{\tau - \tau_0}{\tau};$$

which was to be demonstrated.

112. Another and a much simpler demonstration is readily obtained from equations (75) and (89), which give

$$dQ = \tau d\phi.$$

As τ is constant for isothermal lines, this equation gives for the cycle of Carnot between M and N, the definite integral

$$q = \tau\,(\phi - \phi_0)\,;$$

and similarly, for the isothermal line τ_0 between P and Q, the quantity of heat

$$q_0 = \tau_0 (\phi - \phi_0);$$

while for the adiabatic lines there is no variation of quantity.

Consequently, we have

$$\frac{q - q_0}{q} = \frac{\tau - \tau_0}{\tau_0},$$

for the efficiency of cycles of Carnot.

It is not possible to construct an engine in which changes occur without loss by conduction and radiation; or, in other words, to realize changes by adiabatic lines; yet we may readily reduce to infinitesimal cycles of Carnot the operations of engines generally, and find their work as a definite integral of such elementary cycles.

CARNOT'S THEOREM.

113. It has been announced, without proof, that, for all heat engines whatever, the *duty*, or maximum efficiency, is

$$\frac{q - q_0}{q} = \frac{\tau - \tau_0}{\tau_0};$$

this expression has just been shown to be that of the efficiency of a perfect air engine working in a cycle of Carnot; the same expression has also been obtained for air engines of Stirling and Ericsson, if rendered perfect by means of a regenerator. And it will now be shown to be impossible that any heat engine whatever can have a greater coefficient than that of an air engine working in a cycle of Carnot; which being reversible should therefore be perfect.

When any two engines have equal coefficients of efficiency, we must have

$$\frac{q - q_0}{q} = \frac{q' - q_0'}{q'};$$

consequently,

$$\frac{q}{q'} = \frac{q - q_0}{q' - q_0'} = \frac{q_0}{q_0'}. \qquad (116)$$

If now it be imagined that any other engine whatever, M, can be more efficient; in other words can, with an equal amount of heat, do more work than an air engine, N, which works in a reversible cycle of Carnot; then, let M make m cycles while N makes n cycles, and let q and q' be the quantities of heat respectively emitted and received by the source for a single cycle; while q_0 and q'_0 are those given to and taken from the refrigerator. The limiting temperatures τ and τ_0 being the same for both engines.

The quantity of heat to be used being, by hypothesis, the same for both engines, we must have

$$mq - nq' = 0.$$

Hence, if the two engines be equally efficient,

$$m(q - q_0) - n(q' - q_0') = 0.$$

But if M be more efficient than N, then will this difference, or

$$nq_0' - mq_0,$$

be a positive quantity.

Combine the two engines into one compound machine, M the more efficient driving N reversely. The engine M receives from the source A in m cycles the quantity of heat mq and gives to B, the refrigerator, mq_0. While N worked backwards takes from B the quantity nq_0' and gives to A the quantity nq'.

The source A therefore imparts the total quantity of heat

$$mq - nq';$$

but this is, by hypothesis, zero; and consequently A neither receives nor emits any heat whatever.

The refrigerator B receives mq_0 and gives to A the quantity nq_0'; it therefore imparts

$$nq_0' - mq_0.$$

Now, either the work done must of necessity be zero, in which case the two engines are equally efficient and the last expression is zero; or the work done by M in driving N must be produced by a positive quantity of heat given by B to A, the value of which is determined by the expression just found.

This is simply impossible, for it is contrary not only to all our knowledge and experience, that work should be done by a cold body giving heat to a hot one, but also to the first principles of the dynamical theory; according to which heat is energy, readily imparted with a *chute de chaleur* by a hot to a cold body, by moving to motionless particles; but not capable of the reverse transference from cold to hot bodies with accompanying performance of mechanical work done by it. Such a proposition is the equivalent of supposing ice to generate steam and thereby work an engine; or water to work mills by running up hill.

The principle, that work cannot be done by a cold body imparting heat to a hot one, is a thermodynamic axiom first proposed by Sir W. Thomson. One less obvious was used by Clausius.

114. As it has now been proved that no engine can be more efficient than an air engine working in a cycle of Carnot, it follows that for all engines the *duty*, or maximum, is

$$\frac{q - q_0}{q} = \frac{\tau - \tau_0}{\tau}.$$

This remarkable and very important equation, the same as (109), though first definitely determined by Sir W. Thomson, is usually called *the theorem of Carnot;* for to Carnot we owe the law that "the mechanical power of heat is independent of the agents employed to realize it; its quantity being fixed solely by the temperatures of the bodies between which the heat is transported;" which enunciation may be symbolically written

$$e = C(\tau - \tau_0). \tag{117}$$

Unhappily, Sadi Carnot was so far misled by the then prevalent material hypothesis, that he failed to determine the factor C, which Thomson has since named the *function of Carnot.*

On comparison of the last two equations, it is evident that

$$C = \frac{1}{\tau}, \tag{118}$$

or that *the function of Carnot is the reciprocal of the absolute temperature.*

Conversely, it is sometimes given, as a definition of absolute temperature, that it is the reciprocal of this function.

AIR ENGINES COMPARED WITH STEAM ENGINES.

115. Having found for all heat engines the same *duty,* or coefficient of maximum efficiency, we may now compare the actual working results of air engines, such as that of Ericsson, with those of one of the steam engines of Hirn.

Suppose a perfect air engine to work at the temperatures of one of Hirn's steam engines, for which he observed $t = 146°$ and $t_0 = 34°$. We have also

$$\alpha = \frac{1}{273} = \frac{1}{a}.$$

If we substitute these values we find

$$\frac{\tau - \tau_0}{\tau} = \frac{t - t_0}{a + t} = \frac{112}{419} = \frac{2}{7} \text{ nearly,}$$

which is therefore the maximum.

Now, for his steam engine, Mr. Hirn obtained $\frac{1}{8}$ as its actual efficiency; and it is very probable that an air engine, working between 146° and 34°, would not give one-eighth of the total expenditure of heat as its coefficient.

Experiments made in Paris, at the *Conservatoire des Arts*, under the direction of MM. Morin and Tresca, gave for an air engine of Ericsson the result, that 4.13 kilog. of coke, or 5.85 kilog. of bituminous coal, per horse power, were consumed every hour. The actual disposable work was only 0.27 of that measured by the indicator. The latter coefficient shows very inferior workmanship, for experiments made with an indicator and a *frein de Prony* had given, for the ratio of disposable work to that shown by an indicator, 0.80 to 0.85 in well-constructed steam engines.

Other experiments made with an air engine of M. Laubereau, in his presence and with his aid, gave MM. Morin and Tresca the consumption per horse-power of 4.55 kilog. of fuel per hour.

According to the same able observers, the consumption of bituminous coal per horse-power in steam engines varies from 1.2 to 6 kilog. per hour.

These practical working results do not, therefore, indicate any superiority for air engines; nor has any engine yet proved itself to be, when considered in every respect, more advantageous than the best steam engines.

The most serious attempt to substitute the power of heated air for that of steam was that made at New York, about twenty years ago, with the ship Ericsson, which was at first provided with air engines constructed by Capt. Ericsson. The experiment on so large a scale was instructive, but unsuccessful. And the ship was subsequently altered into an ordinary steamer. Small Ericsson engines are, however, still used in New York, and if not more economical than steam engines, they at least are less dangerous in some respects.

116. An inspection of the formula of maximum efficiency,

$$\frac{\tau - \tau_0}{\tau},$$

shows that advantage is gained either by increasing the numerator or by lessening the denominator of this ratio. The numerator may be increased by reducing τ_0; if it were the absolute zero, the value of the ratio would become unity and the whole of the heat Q received could be utilized by a perfect engine. Upon this result a dynamical definition of the absolute zero has been founded.

But τ_0 is zero only at $-273°$ C., a degree of cold wholly unattainable. If to increase the numerator we elevate the temperature τ, we thereby also augment the denominator and partially neutralize the advantage to be gained.

As limits the freezing point of water, or 0° C., and the temperature of incandescence or redness, about 500° C., may be considered physically attainable or possible in heat engines; these temperatures give

$$\frac{\tau - \tau_0}{\tau} = \frac{500}{773} = 0.64;$$

such therefore is the maximum coefficient for a perfect engine of any kind working with a *chute de chaleur* from redness to the cold of melting ice.

An engine whose faults of construction are so slight as to give a practical or utilizing coefficient of 0.85 of the theoretical work might consequently give 0.64 multiplied by 0.85, or the coefficient

0.54, as the actual efficiency of a very perfect heat engine between such extreme temperatures.

But high temperatures even much below that of redness, producing powerful strains by irregular expansion, and causing oxidation and other deterioration, are so destructive that, even in air engines, it has not been found advantageous to elevate the temperature much above that used in steam engines. And unless it be for the purpose of very greatly extending the *chute de chaleur*, or range of temperature between τ and τ_0, there seems to be no important advantage to be gained by substituting heated air for steam.

CHAPTER VIII.

THERMAL LAWS.

THEOREM OF CARNOT.

117. The law of equivalence of Joule is only the definite expression of the fundamental truth that heat is energy,

$$Eq = \Sigma \int P dp.$$

Not less important is the theorem of Carnot, that it is impossible to employ efficiently, or utilize, of a given quantity of heat q, more than the *chute de chaleur*, or fraction,

$$q \frac{\tau - \tau_0}{\tau}.$$

Hence this theorem is called *the second fundamental law* of thermodynamics. And as quantities of heat are not directly measurable, the law is the more valuable in that it enables us to substitute for them absolute temperatures indicated by an air thermometer.

Moreover, being universally true, the theorem of Carnot can be used to generalize many restricted propositions obtained for gases only, from the law of Charles and Mariotte; and thus we can establish general thermal laws applicable to all bodies.

To this extension and generalization chiefly, attention is invited in this chapter. But before entering upon that subject, it is well to familiarize ourselves with the various equivalent algebraic forms in which the theorem of Carnot is usually expressed.

By the law of Joule, the work of an engine, receiving q from the source and giving q_0 to the refrigerator, without waste, or the total dynamical value of the heat used, is

$$E(q - q_0) = \Sigma \int P dp.$$

But by Carnot's theorem,

$$q - q_0 = \frac{q}{\tau}(\tau - \tau_0);$$

therefore

$$\Sigma \int P dp = E(q - q_0) = E\frac{q}{\tau}(\tau - \tau_0),$$

which is the complete analytical expression of the two laws.

As perfect engines have all the same *duty*,

$$\frac{q - q_0}{q} = \frac{\tau - \tau_0}{\tau} = \frac{q' - q_0'}{q'}; \tag{119}$$

which gives always the constant ratio,

$$\frac{\tau}{\tau_0} = \frac{q}{q_0} = \frac{q'}{q_0'} = \frac{q''}{q_0''} = \text{etc.}; \tag{120}$$

this result may be thus enunciated: between the same temperatures, the ratio of the quantity of heat given to the refrigerator to the total quantity received by any substance, or system, is constant for all bodies working in cycles of Carnot; and is equal to the ratio of the corresponding temperatures.

Equation (119) may be written thus:

$$\frac{\Delta\tau}{\tau} = \frac{\Delta q}{q} = \frac{\Delta q'}{q'} = \frac{\Delta q''}{q''} = \text{etc.}$$

And as this is true for all values, it is true when the variation is infinitesimal, or

$$\frac{d\tau}{\tau} = \frac{dq}{q} = \frac{dq'}{q'} = \frac{dq''}{q''} = \text{etc.} \tag{121}$$

The preceding expressions give also the constant ratio,

$$\frac{q}{\tau} = \frac{q_0}{\tau_0} = \frac{q'}{\tau'} = \frac{\Delta q}{\Delta \tau} = \frac{dq}{d\tau}; \tag{122}$$

to which constant ratio (122) Clausius gives the rather obscure name of "*equivalence-value.*"

It is evident that these different forms simply express that, not only for perfect gases, but for all substances, absolute temperatures vary proportionally to the quantities of heat absorbed and given out in cycles of Carnot, or in perfect engines. From equation (119) we obtain, for the *heat necessarily lost in perfect engines,*

$$\frac{\tau_0}{\tau} = \frac{q_0}{q} = \frac{q_0'}{q'} = \frac{q_0''}{q''} = \text{etc.} \tag{123}$$

To express which in ordinary temperatures, we have

$$\frac{\tau_0}{\tau} = \frac{a + t_0}{a + t} = \frac{1 + \alpha t_0}{1 + \alpha t};$$

and for the duty

$$\frac{\tau - \tau_0}{\tau} = \frac{t - t_0}{a + t}. \tag{124}$$

By the law of Joule, Edq is the total dynamical value, or equivalent, of the variation of heat dq; but by the theorem of Carnot the actual value, or proportion, of this heat which can be used in an engine is only

$$E \frac{\tau - \tau_0}{\tau} dq.$$

If now in any cycle, q be the heat received and q_0 that emitted, then will

$$E(q - q_0) - E\tau_0 \int \frac{dq}{\tau}$$

be the work, during the cycle, of the heat which can possibly be utilized.

But if the engine be perfect, or the cycle be reversible, the first term denotes the amount of work; and the second term consequently becomes

$$\int \frac{dQ}{\tau} = 0. \tag{125}$$

Generally, however, as engines are imperfect, and therefore not reversible, the work during a cycle is much less than the value of the first term, which expresses its amount in perfect engines only; consequently,

$$\tau_0 \int \frac{dQ}{\tau} \tag{126}$$

is the heat wasted, and neither converted into useful work, nor necessarily given to the refrigerator, as q_0 must always be. Hence the expression just found is called by Thomson, who first obtained it, that of the *dissipation;* and it measures the imperfection, which cannot possibly be a negative quantity.

GENERAL EQUATION FOR ALL TRANSFORMATIONS.

118. We have proved that there is always a factor capable of rendering exact and integrable the partial differential equations of thermodynamic changes. We have also found that, for perfect gases, this factor is

$$\lambda = a + t = \tau,$$

the absolute temperature, as defined and indicated by an air thermometer. So that our general equation of transformation for all substances,

$$dQ = \lambda d\phi,$$

becomes for perfect gases

$$dQ = \tau d\phi. \tag{127}$$

The theorem of Carnot serves to generalize this result, by proving that for all bodies λ is equal to τ, the absolute temperature; while ϕ is a determinate but unknown function for each particular substance; the form of which can be obtained for perfect gases only.

119. To show that the factor λ is equal to τ for all bodies, let an engine work in a cycle of Carnot, $MNPQ$, composed of two isothermal lines, τ and τ', and two adiabatic lines, ϕ and ϕ'; which may be taken so near to each other that

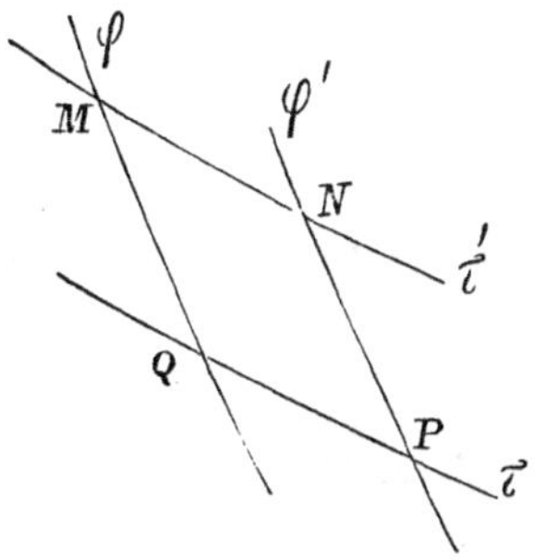

$$\phi' = \phi + d\phi.$$

Let λ' and λ denote the values of λ at M and Q; then the general equation of transformation (75) gives for the changes from M to N, and from P to Q,

$$dq' = \lambda' d\phi,$$

$$dq = \lambda \, d\phi;$$

from which and equation (122) we get

$$\frac{\lambda'}{\lambda} = \frac{q'}{q} = \frac{\tau'}{\tau}.$$

But the ratio of τ to τ', or that of q to q', has been proved to be constant for all bodies; it follows, therefore, that such is necessarily the case for the equal ratio of λ to λ'. Consequently, the factors λ and λ' must equal the same function of τ and τ', multiplied by an arbitrary function u, dependent in each case upon the nature of the body, or algebraically,

$$\frac{\lambda'}{\lambda} = \frac{u\, f(\tau')}{u\, f(\tau)} = \frac{\tau'}{\tau}.$$

The function u being arbitrary, and therefore capable of an infinite number of values, see § 62, we may put u equal to unity, which gives for all bodies

$$\frac{\lambda'}{\lambda} = \frac{f(\tau')}{f(\tau)} = \frac{\tau'}{\tau}.$$

Now it has been found, for perfect gases, that λ is equal to the absolute temperature τ, consequently for all bodies

$$\begin{aligned} \lambda &= f(\tau) = \tau, \\ \lambda' &= f(\tau') = \tau'; \end{aligned} \tag{128}$$

and, therefore, we have always

$$dQ = \tau d\phi; \tag{129}$$

in which equation ϕ is a determinate function of the independent variables, but differing with the nature of the substance.

This equation is perfectly general and applicable to all thermodynamic changes. Hence, it is called by Rankine the *general thermodynamic function.*

120. Integrating (129) between limits, and supposing τ constant, or the change to occur isothermally, we get

$$Q_1 - Q_0 = \tau(\phi_1 - \phi_0); \tag{130}$$

or the quantity of heat requisite for any body to pass from an adiabatic line ϕ_0 to another ϕ', by an isothermal change, is proportional to the temperature.

From this result, we may readily get the expression for the theorem of Carnot. Let τ and τ_0 be the temperatures of the isothermal lines, and ϕ and ϕ_0 be the adiabatic lines of a cycle of Carnot, then

$$q = \tau(\phi - \phi_0),$$

$$q_0 = \tau_0(\phi - \phi_0),$$

and

$$\frac{q}{\tau} = \frac{q_0}{\tau_0};$$

consequently,

$$\frac{q - q_0}{q} = \frac{\tau - \tau_0}{\tau};$$

which is Carnot's theorem of maximum efficiency, and true for all substances.

121. Integrating between the limits (1) and (2), we obtain

$$Q_2 - Q_1 = \int^2 \tau d\phi;$$

the second member of which depends not only upon the initial and final states (1) and (2) but upon all the intermediate states. To render this evident, we have by equation (58), the dynamical result

$$E\int_1^2 \tau d\phi = U_2 - U_1 + \int_1^2 pdv.$$

If now in the annexed diagram we suppose a body to pass from the state M to the state N, it is perfectly clear that the second term of the second member, or the integral of pdv, will be represented by $M'MNN'$, and that this area is a function of all the consecutive intermediate values of p between M and N. Moreover, its value for a complete cycle is evidently represented by the closed area included in the curve or diagram of energy; but that of the internal energy is zero.

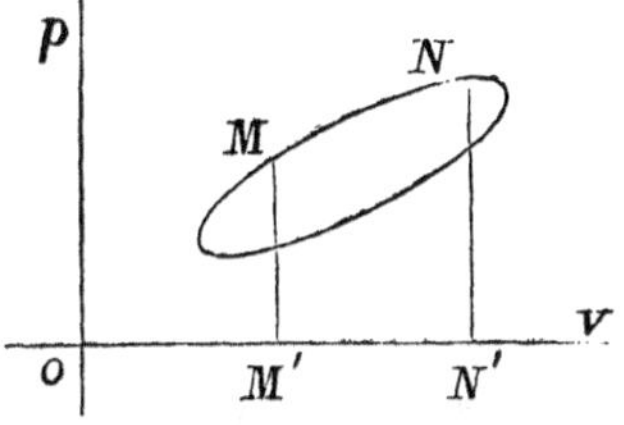

122. To prove from equation (129) that the efficiency of the cycle of Carnot is the maximum; let that of any other cycle be represented by the closed area *abcd* of the annexed figure. Now, through its points *b* and *d* of highest and lowest temperature, τ and τ', and the extreme points *a* and *c*, corresponding to the greatest and least values ϕ and ϕ' of the function ϕ, we may always draw isothermal and adiabatic lines, forming a cycle of Carnot tangential at *a*, *b*, *c*, *d* to the given cycle.

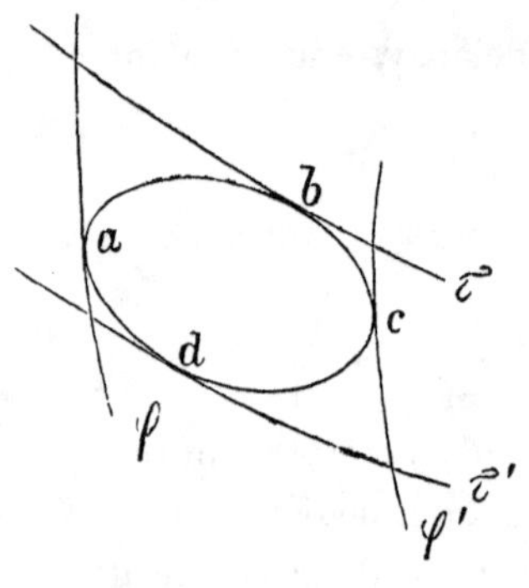

It is evident, upon mere inspection, that the area *abcd* representing the external work is less than that of the circumscribed cycle of Carnot. But it may be imagined that through the extreme points *a*, *b*, *c*, *d* other lines of transformation may be drawn which would give a greater area or efficiency. By hypothesis, τ and τ' are the extremes of temperature; no line passing through the points *b* and *d* can therefore give for

$$q = \int \tau d\phi$$

so great a value as the isothermal line for which the factor τ is the maximum and constant. And no line passing through *d* can give for

$$q' = \int \tau' d\phi$$

so small a value as the isothermal line for which the tenperature τ' is the minimum.

It is also evident that the definite integral of $d\phi$ in these equations is a maximum when ϕ and ϕ' are constant and the tangential lines through *a* and *c* are consequently adiabatic.

Performing the integrations indicated in the preceding discussion, and reducing, we obtain for the efficiency of the cycle of Carnot, thus proved to be a maximum,

$$\frac{q - q_0}{q} = \frac{\tau - \tau_0}{\tau}.$$

And thus again it is shown to be impossible, if the equations

$$dq = \tau d\phi,$$
$$dq' = \tau' d\phi,$$

represent the heat received and emitted by any substance, that an engine can ever utilize of that received q more than

$$q - q' = q\frac{\tau - \tau_0}{\tau}; \qquad (131)$$

in which τ and τ_0 are the temperatures of the source and the refrigerator.

EQUATIONS OF SIR W. THOMSON.

123. By the general thermodynamic function

$$\frac{dQ}{\tau} = d\phi,$$

but v and τ being the independent variables,

$$\frac{dQ}{\tau} = c\frac{d\tau}{\tau} + l\frac{dv}{\tau};$$

and this is an exact differential $d\phi$; consequently,

$$l = \tau\left(\frac{dl}{d\tau} - \frac{dc}{dv}\right); \qquad (132)$$

which by virtue of (73) reduces to

$$l = A\tau\frac{dp}{d\tau}, \qquad (133)$$

identical with

$$\frac{dp}{d\tau} = E\frac{l}{\tau}. \qquad (134)$$

This is evidently true for any substance or transformation whatever in which p varies as a function of v and τ.

In like manner, from the equation

$$\frac{dQ}{\tau} = c'\frac{d\tau}{\tau} + h\frac{dp}{\tau},$$

$$\tau\frac{dc'}{dp} = \tau\frac{dh}{d\tau} - h,$$

and

$$h = \tau\left(\frac{dh}{d\tau} - \frac{dc'}{dp}\right); \tag{135}$$

which, by virtue of equation (74), reduces to

$$h = -A\tau\frac{dv}{d\tau}. \tag{136}$$

Taking now the specific volume v and the pressure p for independent variables,

$$\frac{dQ}{\tau} = M\frac{dv}{\tau} + N\frac{dp}{\tau},$$

consequently,

$$\tau\frac{dM}{dp} - M\frac{d\tau}{dp} = \tau\frac{dN}{dv} - N\frac{d\tau}{dv};$$

therefore,

$$M\frac{d\tau}{dp} - N\frac{d\tau}{dv} = \tau\left(\frac{dM}{dp} - \frac{dN}{dv}\right).$$

And this, by virtue of equation (72), reduces to

$$M\frac{d\tau}{dp} - N\frac{d\tau}{dv} = A\tau \tag{137}$$

124. Equation (137) contains both of the derivatives

$$\frac{d\tau}{dp} \quad \text{and} \quad \frac{d\tau}{dv};$$

but equations (133) and (136) contain each only one of them.

If we denote the general thermodynamic function by

$$\phi(p, v, t) = 0,$$

and make v constant, then we have

$$\frac{d\tau}{dp} \cdot \frac{dp}{d\tau} = 1;$$

and, therefore, equation (133) is equivalent to

$$l\frac{d\tau}{dp} = A\tau. \tag{138}$$

Similarly, by making p constant,

$$\frac{d\tau}{dv} \cdot \frac{dv}{d\tau} = 1,$$

and equation (136) may be transformed into

$$h\frac{d\tau}{dv} = -A\tau. \tag{139}$$

Hence it follows that the same function τ of p and v satisfies all three of the equations (137, 138, 139) derived from the general equation of transformation.

125. For an isothermal change, Sir W. Thomson gives the following mode of determination: "let the substance expand from any volume v_0 to v, and being kept constantly at the same temperature τ, let it absorb the quantity q of heat. Then

$$q = \int_{v_0}^{v} l dv = A\tau \int_{v_0}^{v} \frac{dp}{d\tau} dv.$$

"But if w denote the mechanical work which the substance does in expanding, we have

$$w = \int_{v_0}^{v} p dv,$$

and therefore

$$q = A\tau \frac{dw}{d\tau}. \qquad (140)$$

"This formula, established without any assumption admitting of doubt, expresses the relation between the heat developed by the compression of any substance whatever, and the mechanical work which is required to effect the compression, as far as it can be determined without hypothesis, by purely thermal considerations."

EQUATIONS OF RANKINE.

126. If in the general thermodynamic equation,

$$d\phi = c\frac{d\tau}{\tau} + l\frac{dv}{\tau},$$

we substitute for l the value given in equation (133) and integrate, we obtain

$$\phi = c \log \tau + A \int \frac{dp}{d\tau} dv. \qquad (141)$$

And similarly, by integration and substitution of the value (136) for h, in the equation

$$d\phi = c'\frac{d\tau}{\tau} + h\frac{dp}{\tau},$$

it gives

$$\phi = c' \log \tau - A \int \frac{dv}{d\tau} dp. \qquad (142)$$

To separate the second member of (141) into terms which shall indicate the internal potential energy of molecular action and the work done against external pressure; we have for the total differential of ϕ, considered as a function of v and τ,

$$d\phi = \frac{d\phi}{d\tau} d\tau + \frac{d\phi}{dv} dv.$$

Hence, equation (141) gives, by differentiation,

$$\tau d\phi = c d\tau + A\tau \int \frac{d^2p}{d\tau^2} dv \cdot d\tau + A\tau \frac{dp}{d\tau} dv. \qquad (143)$$

This equation is much used by Rankine, who analyzes it as follows: the term $cd\tau$ is the energy which the body possesses in virtue of being *hot;* the second term

$$A\tau \int \frac{d^2p}{d\tau^2} dv \cdot d\tau$$

is the heat which produces work by mutual molecular actions dependent upon the temperature; and the last term

$$A\tau \frac{dp}{d\tau} dv$$

is the heat of expansion performing external work, as well as working internally against molecular attraction dependent upon variation of volume.

Another analysis by Rankine is: 1°, the change of sensible heat $cd\tau$ as before; 2°, the external work pdv, represented by a diagram of energy; 3°, the internal work performed in overcoming molecular action

$$\tau \int \frac{d^2p}{d\tau^2} dv . d\tau + \left(\tau \frac{dp}{d\tau} - p\right) dv. \qquad (144)$$

127. For a body in the perfectly gaseous state, we have found

$$l = Ap;$$

therefore, by equation (133), this gives

$$p = \tau \frac{dp}{d\tau};$$

and from this, by differentiation, we find

$$\frac{dp}{d\tau}d\tau = \tau\frac{d^2p}{d\tau^2}d\tau + \frac{dp}{d\tau}d\tau\,;$$

consequently,

$$\tau\frac{d^2p}{d\tau^2}d\tau = 0. \tag{145}$$

Both terms of the expression (144) vanish, and the internal molecular work is therefore zero. Heat is consequently entirely consumed in rendering the body hotter and in performing external work.

Writing the law of Charles and Mariotte under the form

$$pv = \frac{p_0v_0}{\tau_0}\tau = R\tau, \tag{146}$$

and differentiating p as a function of τ, we have

$$\frac{dp}{d\tau} = \frac{R}{v} = \frac{p}{\tau}\,; \tag{147}$$

and thus again we arrive at the expression used above.

It has been shown, equation (80), that for perfect gases

$$c' - c = A\,\alpha\,p_0v_0\,;$$

which, when p_0, v_0, τ_0 are assumed to be the pressure, volume, and temperature for the state of melting ice, becomes

$$c' = c + A\frac{p_0v_0}{\tau_0}. \tag{148}$$

By substituting this value of c' in equation (142), we obtain

$$\phi = \left(c + A\frac{p_0v_0}{\tau_0}\right)\log\tau - A\int\frac{dv}{d\tau}dp. \tag{149}$$

The specific heat of constant pressure c' is evidently equal to the specific heat of constant volume c, increased by the quantity of heat required for an expansion corresponding to one thermometric degree, in an unit of weight of the gas.

If we differentiate equation (149) as a function of p and τ, we readily obtain

$$\tau d\phi = \left(c + A\,\frac{p_0 v_0}{\tau_0} - \tau A\int\frac{d^2v}{d\tau^2}\,dp\right)d\tau - A\tau\,\frac{dv}{d\tau}\,dp\,; \qquad (150)$$

another equation of Rankine analogous to (143) and of similar use.

In the expressions (143) and (150), the factors containing the second derivatives

$$\frac{d^2p}{d\tau^2} \quad \text{and} \quad \frac{d^2v}{d\tau^2}$$

represent the deviation of the gas from the laws of Charles and Mariotte, or from the perfectly gaseous state.

128. Another form of the general thermodynamic function used by Rankine is

$$dQ = \tau d\phi = c d\tau + \tau dF\,; \qquad (151)$$

in which F is called by him the "*metamorphic function*," or the "*heat potential.*" Comparing this with equation (143), it is evident that

$$dF = A\int\frac{d^2p}{d\tau^2}\,dv.d\tau + A\,\frac{dp}{d\tau}\,dv\,;$$

whence, by integration,

$$F = A\int\frac{dp}{d\tau}\,dv.$$

But if the total work, internal and external, be denoted by

$$w = \int_{v_0}^{v} p dv,$$

$$\frac{dw}{d\tau} = \int\frac{dp}{d\tau}\,dv,$$

and

$$F = A\frac{dw}{d\tau};$$

consequently,

$$Fd\tau = Adw \tag{152}$$

is the heat equivalent of the elementary work dw, or the amount of heat required for its performance, independently of the specific heat of constant volume which causes the variation of temperature.

SIMPLE ENUNCIATION BY RANKINE OF THE SECOND LAW.

129. As the second law, algebraically expressed in the form

$$\frac{\Delta q}{q} = \frac{\Delta\tau}{\tau} = \frac{E\Delta q}{Eq},$$

merely asserts, for absolute temperatures, their proportionality to the corresponding quantities of heat absorbed and emitted in perfect engines, it is evident that this truth may, conversely, be so stated as to constitute an expression of the second fundamental law.

The work of q being Eq, and Δq and $E\Delta q$ being their similar submultiple or fractional parts, the above expression may be written as follows:

$$n\Delta q = q; \qquad n\Delta\tau = \tau; \qquad nE\Delta q = Eq;$$

or thus

$$\Sigma\Delta q = q; \qquad \Sigma\Delta\tau = \tau; \qquad \Sigma E\Delta q = Eq.$$

Hence, for the second general law, we have the following enunciation: if the total heat q, or absolute temperature τ, of any uniformly heated mass be subdivided into any number of equal parts, the energy or work will be the sum of the equal effects of those equal parts.

This is only a particular case of the more general proposition of Rankine in 1853, that, in causing transformations, the effect of a quantity of energy is the sum of the effects of all its parts.

For this proposition, Rankine gives the following graphic construction. Let $ABCD$ be a diagram of energy between the

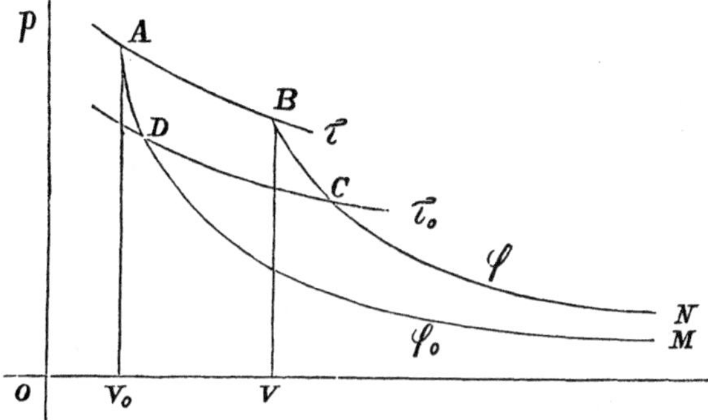

two isothermal lines τ and τ_0, and the two adiabatic lines ϕ and ϕ_0, indefinitely prolonged. Let $\Delta\tau$ be the difference between τ and τ_0, and such that $n\Delta\tau$ is equal to τ. Then will the area $ABCD$ bear to the indefinitely prolonged area $MABN$ the same ratio that $\Delta\tau$ does to τ. Also, this area $ABCD$ represents the transformation of heat into work represented by the abstraction of any one of the equal parts $\Delta\tau$ into which τ is divided, and the effect of τ is the sum of the effects of its parts $n\Delta\tau$.

For this theorem Rankine also gives the following symbolical exposition. Let the temperature vary by $\delta\tau$, then will the pressure vary by

$$\frac{dp}{d\tau}\,\delta\tau,$$

and the quadrilateral $ABCD$ will be expressed by

$$\delta\tau\int_{v_0}^{v}\frac{dp}{d\tau}\,dv.$$

consequently, the indefinite or total area *MABN*, or the latent heat of expansion, gives

$$\int ldv = A\tau \int \frac{dp}{d\tau} dv\,;$$

which is evidently identical with equation (133) already obtained.

GENERALIZATION OF ABSOLUTE TEMPERATURES.

130. For the function of Carnot, we have the expression

$$c = \frac{1}{\tau}\,;$$

giving for all substances the definition, that absolute temperature is the reciprocal of the function of Carnot.

We have also proved, first for the perfectly gaseous state, and subsequently for all bodies, equation (128), that the factor of integrability λ, which renders exact the partial differential equations of thermal transformation, is simply τ the absolute temperature. And the laws of perfect gases establish that, in the equation

$$\lambda = a + t = \tau, \tag{153}$$

the temperature t is that indicated by an air thermometer; and a is the reciprocal of the coefficient of dilatation for a gas obeying perfectly the laws of Charles and Mariotte; the value of which does not differ much from

$$a = -\,273^\circ,$$

according to the usual centigrade scale.

These results are of the greatest value, for they give for the general thermodynamic function the form

$$dQ = \tau d\phi\,;$$

in which the factor λ, having been replaced by τ, is no longer of indeterminate signification or value.

Moreover, by eliminating the arbitrary indications of common thermometers, and substituting for them absolute temperatures dependent upon the nature of heat itself, and which do not vary with the thermometric substance employed, we introduce into thermodynamic expressions that generality and clearness which belongs to the laws of Nature, ever comprehensive, simple and beautiful when clearly understood. But this simplicity vanishes if we employ such arbitrary thermometers as those of Fahrenheit, Reaumur and Celsius, made capriciously to depend upon the relative expansion of mercury and glass, and the particular temperatures of melting ice and boiling water.

Yet, as thermometric observations are nearly all made with ordinary thermometers, equation (153) is of great value to convert ordinary into absolute temperatures; provided that we employ air thermometers, or reduce the indications of common mercurial thermometers to corresponding degrees of the air thermometer by applying the requisite corrections.

131. It is necessary that we now seek to generalize the definition of absolute temperatures. Heat being due to motion, it is evident that, as already stated in § 70, rest or the absence of motion will give the absolute zero of temperatures. But to determine this zero and an absolute scale, we have used the laws of Charles and Mariotte. This seems inadmissible, in so far as absolute temperatures are thus made to depend upon the expansion of perfect gases only. The defect has been removed by Sir W. Thomson, by a happy generalization of the definition of absolute temperature, so as to make it the same for all bodies. In the law of efficiency

$$\frac{q - q_0}{q} = \frac{\tau - \tau_0}{\tau},$$

the temperatures τ and τ_0 are understood to be such as are indicated by air thermometers; yet this law is true for all substances whatever. So also, therefore, is the law

$$\frac{q}{q_0} = \frac{\tau}{\tau_0};$$

hence, absolute temperatures may be defined to be such as are proportional to the quantities of heat received and emitted in perfect engines, or in cycles of Carnot.

In the language of Thomson: "the temperatures of two bodies are proportional to the quantities of heat respectively taken in and given out in localities at one temperature and at the other respectively, by a material system subjected to a complete cycle of perfectly reversible thermodynamic operations, and not allowed to part with or take in heat at any other temperature. Or the absolute values of two temperatures are to one another in the proportion of the heat taken in to the heat rejected in a perfect thermodynamic engine, working with a source and a refrigerator at the higher and lower of the temperatures respectively."

This definition, thus made to flow from the fundamental law of efficiency, is evidently perfectly general and independent of the substance employed; while it accords with and includes that deduced from the laws of Charles and Mariotte.

132. If we make τ_0 equal to zero, in the equation of efficiency

$$\frac{\tau - \tau_0}{\tau} = \frac{q - q_0}{q},$$

then q_0 becomes zero, and the heat q received from the source is entirely converted into work. Hence the absolute zero is defined to be *that value of* τ_0 *which would cause the whole of the heat to be utilized.*

133. It should be remembered that the word *temperature* is simply the name used to indicate the relative state of one body to another when the colder receives heat from the hotter. And that the measurement of temperatures by expansion in thermometers, based upon the arbitrary assumption that variations of volume are *proportional* to changes of temperature, is, as we have fully shown, not true, even approximately, except for thermometers made of permanent gases. Such erroneous assumptions may have answered their purposes two centuries ago; but that volumes vary proportionally to temperature is now no longer a postulate, which may be conceded as a convenient basis for a faulty definition, but a proposition to be refuted or verified for any substance by exact experimental investigation.

A transfer of heat from a hot to a colder body is thus expressed,

$$dq = cd\tau + ldv,$$

the first term of the second member denoting the change of temperature, and the last that of volume. These changes are evidently perfectly distinct.

For clearness of conception, we need to bear in mind that sensible heat, light, and sound are all effects of vibratory motion; and as the variations in the physiological sensation which we call differences of brilliancy and colour for light, and of loudness and pitch for sound, depend upon the *vis viva*, or the maximum displacement and time of vibration, so are changes of temperature, or hotness, analogous functions of the molecular vibrations in any substance.

HEAT MEASURED DYNAMICALLY.

134. As heat is *energy*, or power to do work, it is clear that quantities of heat, instead of being estimated in thermal units, may be measured by the proportional dynamical work which they can perform. For this purpose, we have the equation

$$\varepsilon = \int Pdp = Eq, \tag{154}$$

in which E is Joule's factor, and ε the work or energy corresponding to q, the number of thermal units or quantity of heat.

From equation (154) we readily obtain, for perfect engines,

$$\frac{\varepsilon - \varepsilon_0}{\varepsilon} = \frac{\tau - \tau_0}{\tau}; \tag{155}$$

another expression for Carnot's theorem, in which ε and ε_0 are units of work, or kilogrammetres.

This simple dynamical mode of measuring heat is often preferable, and is much used, especially by Rankine in his book on the steam-engine.

DIFFERENT FORMS OF STATEMENT OF THE FUNDAMENTAL LAWS.

135. The variety of forms used to express the fundamental laws may slightly perplex a beginner. It is not sufficient to state the first law thus simply,

$$Eq = \Sigma \int Pdp,$$

for internal work must be eliminated.

Hence Clausius, instead of using the equation

$$Eq = U + S, \tag{46}$$

finds it necessary, in all applications, to employ

$$dq = AdU + Apdv, \tag{71}$$

eliminating U in definite integration by Carnot's principle of restoration to the initial state. In his later writings, Clausius uses the expression

$$dq = dH + \tau dZ,$$

denoting by Z what he calls *disgregation.*

Rankine employs the formula

$$dq = cd\tau + \tau dF, \tag{151}$$

giving to F, the *metamorphic* function, a signification different from that attached by Clausius to Z, his *disgregation.*

As a perfectly general and comprehensive statement of the first fundamental law, Sir W. Thomson gives the equation

$$\frac{dp}{d\tau} = E\left(\frac{dl}{d\tau} - \frac{dc}{dv}\right), \tag{73}$$

obtained from

$$(p - El)\, dv - Ecd\tau,$$

made equal to zero by restoration of the system to its initial state.

136. For the second fundamental law, Rankine uses the function

$$\phi = \int \frac{dq}{\tau}. \tag{129}$$

But Clausius employs in his later writings

$$\frac{q}{\tau} - \frac{q_0}{\tau_0} = 0, \tag{122}$$

putting it under the form

$$\int \frac{dq}{\tau} = 0, \tag{125}$$

and calling it *the principle of the equivalence of transformations.* These formulas had been previously given by Sir W. Thomson, but they are used by Clausius in a manner quite his own.

Both Clausius and Thomson also employ for the second law the equation

$$l = A\tau \frac{dp}{d\tau} \tag{133}$$

demonstrating it differently. And the familiar expression of Carnot's theorem,

$$\frac{q - q_0}{q} = \frac{\tau - \tau_0}{\tau}, \tag{119}$$

is freely used by all these eminent men.

This brief comparative summary is given with the hope that it may aid in the study of their original memoirs, to which the scientific world is indebted for the investigation of these subjects, and all must refer who desire thorough information.

PART II.

APPLICATIONS OF THERMAL LAWS.

CHAPTER IX.

INTRODUCTION.

137. Having discussed the general dynamic laws of heat and the formulas which express them, we propose now to consider some of their more important applications.

All thermal phenomena naturally divide themselves into two classes, those of *internal*, and those of *external* energy. The latter, being accessible to observation, are already quite well known; but the former, with exceptions only, remain hidden and enveloped in mystery.

These exceptions are, however, daily becoming more numerous; and there are few fields of physical discovery more important, or promising, than this difficult one of the internal energy and constitution of bodies.

We are now familiar with the equations

$$EQ = U + S$$

and

$$dQ = AdU + AdS;$$

which express this classification, and in which U denotes *internal* and S *external* work or energy.

138. But before proceeding to apply formulas, it may not be amiss to remark upon the method we have adopted to establish our two general fundamental laws, that of Joule, usually called the first, and that of Carnot, known as the second.

You cannot have failed to perceive that we have simply followed, link by link, and in historic order, the chain of physical discovery; thus presenting our two laws, not as mere mathematical theorems, but as examples of the inductive, or Baconian, method of investigation. To sift evidence, each proposition, or observation, is first challenged and then most severely scrutinized before its admission. Lastly, from the facts a general law embracing the whole group synthetically, and expressive of the relation between them, is inferred; and from this law deduction leads to its consequences. Almost always, however, the law is first only imperfectly reached from a few facts, and then is not considered more than hypothesis, to be confirmed or refuted by extended investigation.

This inductive method, though slow and tedious, demanding patient labour, is yet the only true path of physical discovery,—the path of Galileo, of Newton, of Lavoisier and of Fresnel.

The abstract mathematician, familiar with pithy demonstrations in ancient Greek geometry and with short algebraic processes, becomes impatient of tediousness, and fancies simple and comprehensive ways of reaching his conclusions. But to the physical discoverer all is darkness and night, until glimpses of dawn become harbingers of approaching day.

Not as a mathematical proposition, to be ended with *quod erat demonstrandum*, was the law of celestial gravitation discovered by Newton; nor have other astronomers so found its verification.

And the grand law of all physical energy, or power, that though convertible into many varied forms it is for them all one and indestructible, God alone being able to annihilate what He created, never will or can be proved by algebra, but must rest solely upon its true basis, that of a wide physical induction. The general acceptance of which law is justly regarded as the most important step made in the progress of physical science during this century; a step in which the dynamical theories of heat and light have played no minor part.

We may, with Bartlett, express that universal law by the formula

$$\Sigma\, P\delta p - \Sigma m \frac{d^2 s}{dt^2}\delta s = 0,$$

or with Lagrange by

$$\Pi + V = C,$$

calling it conservation of energy, or of *vis viva;* and for a certain range of purely mechanical truths we may even deduce these equations from the equality of action and reaction, or from the expenditure of power in performance of work. But when we declare them to be the general expression of the mutual transformation and convertibility of any and every kind of physical power, it is clear that this truth is simply incapable of mathematical demonstration, and can only be founded upon induction.

Those habituated imperfectly to algebraic discussion and to geometric reasoning are so accustomed to find conclusions correct, which may have been reached thereby, that they often fall into the error of mistaking shadow for substance, and falsely imagine any result proved to be true, if merely so attained. Nothing can be more fallacious, for diagrams are only auxiliary pictures, and algebra is but a language of limited extent, invented to express relations of quantity and position; whose equations are sentences, composed of verbs, adjectives and substantives; and whose rules are none other than those of universal grammar and correct logic. Nor are there two kinds of logic, one for ordinary thought and words, the other for algebraic expressions;

in both of which, and with almost equal facility, sophistry may be adduced and falsehood perpetrated.

Only upon those clear perceptions of truth, with which God, in his goodness, has endowed human intellect, can we ever found convictions which we may truly call knowledge.

What, therefore, we should require from the algebraic expression of a fundamental physical law is not demonstration or proof thereof, for then the law would be derivative and not fundamental, but simply and solely that it express, or enunciate, the law in the most general, fit and comprehensive manner.

139. Hence, we naturally seek, in an advanced state of knowledge, such expressions for general fundamental laws as are most clear and suitable for our purposes in the use thereof.

The one general formula for all thermodynamic changes

$$dQ = \tau d\phi,$$

so much used by Rankine, is algebraically excellent, and from it we readily deduce many consequences. Thus, for intance, the law of Carnot

$$\frac{\Delta Q}{Q} = \frac{\Delta \tau}{\tau},$$

or the proportionality of absolute temperatures τ to quantities of heat Q, in cycles of Carnot, flows from it directly by definite integration, as has been shown in article 120

But that fundamental formula is objectionable in the feature that it contains the indeterminate or arbitrary function ϕ, for which we know the form only in the hypothetical case of perfect gases.

For the law of Joule we have no better formula than that of Clausius, adopted now by all,

$$dQ = AdU + AdS.$$

And for the law of Carnot there are none better than

$$\frac{Q - Q_0}{Q} = \frac{\tau - \tau_0}{\tau},$$

and

$$l = A\tau \frac{dp}{d\tau};$$

which is the formula of Thomson, presented (in article 123) as a consequence of the law

$$dQ = \tau d\phi$$

and of the partial differential equation (73) of Clausius,

$$A\frac{dp}{dt} = \left(\frac{dl}{dt} - \frac{dc}{dv}\right).$$

In article (129), following Rankine, we have given a simple geometric construction and demonstration of the proposition

$$ldv = A\tau \frac{dp}{dt} dv.$$

That mode of demonstration is due to Clapeyron, who first taught us how to discuss and use cycles of Carnot; and, in modified form, we may here present it as follows:

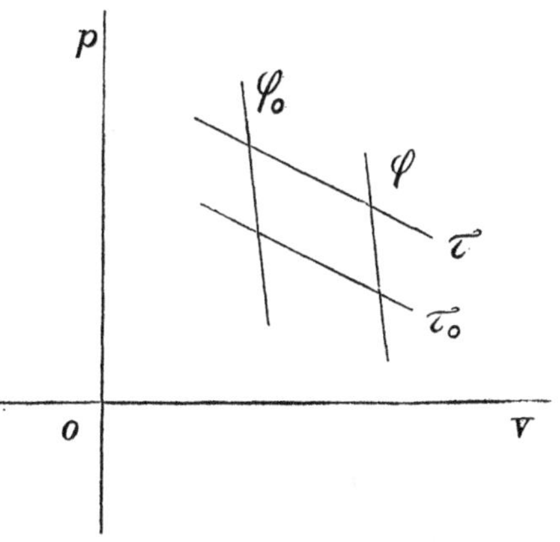

In a diagram of energy for an elementary cycle of Carnot, the infinitesimal quadrilateral area bounded by isothermal and adiabatic lines is measured by

$$\delta p \cdot \delta v;$$

but as the pressure is a function of the temperature

$$\delta p = \frac{dp}{dt}\delta\tau;$$

hence, that area is equal to

$$\delta v \frac{dp}{dt}\delta\tau;$$

and as it measures the mechanical work done, we have only to multiply it by the proper factor to convert it into heat. The factor was unknown to Clapeyron, but is evidently A, the reciprocal of Joule's coefficient.

We have, therefore, for the heat due to this elementary cycle,

$$\delta q = A\delta v \frac{dp}{dt}\delta\tau.$$

To find that due to the isothermal change at τ from ϕ_0 to ϕ, we have

$$\frac{\delta q}{q} = \frac{\delta q}{l\delta v} = \frac{A}{l}\delta\tau \cdot \frac{dp}{dt};$$

whence

$$l = A\frac{q}{\delta q}\delta\tau \cdot \frac{dp}{dt};$$

which Clapeyron put under the simpler form

$$l = C\frac{dp}{dt};$$

but of the factor C he could only determine that it must be a function of the temperature and the same for all bodies.

Beyond this he was unable to go, for with Carnot he ascribed mechanical work done by heat to an imaginary *chute de chaleur*, rather than to the transformation of one kind of energy into another. The material hypothesis demands that δq in a *chute de chaleur* be zero, which renders l infinite. It was, therefore, impossible for Carnot or Clapeyron to do more.

But the labours of Clausius, Thomson and Rankine have since proved that for cycles of Carnot

$$\delta q : q :: \delta\tau : \tau;$$

which gives

$$\tau = \frac{q}{\delta q}\delta\tau,$$

whence

$$= A\tau\frac{dp}{dt}.$$

We have seen how difficult it is in the science of heat to define what is meant by temperature, otherwise than by ascribing it to the *vis viva* of vibrating particles, and regarding it as a state of motion. As ordinarily employed, the word implies a certain condition of the mercury in a thermometer, indicated by and varying with its volume. Evidently, as any definition is arbitrary, we may adopt the proportionality of absolute temperatures τ to their corresponding quantities of heat q, absorbed or emitted in cycles of Carnot, not merely as a physical law, but rather as the very definition itself of what are called *absolute temperatures*. From this definition, then, based on induction, will flow not only Carnot's theorem, as its enunciation, with all the important consequences and applications thereof, but also the valuable formula just established for the relation between the heat expended, or developed, in the mechanical work of expansion, or compression, in any and all substances whatever.

We may evidently put that formula for the latent heat of expansion into the shape of the definite integral

$$\int_{v_0}^{v} ldv = A\tau(v - v_0)\frac{dp}{dt}, \qquad (156)$$

which we shall call the equation of Clapeyron; who first gave it, though in the modified form,

$$Q = C(v - v_0)\frac{dp}{dt};$$

denoting by C an unknown function, identical for all bodies, but since determined; and who also not only gave it, but pointed out clearly its great utility and important applications.

The reader who would entertain just views of the history of our subject should remember always that the labours of Carnot and of Clapeyron constituted, for Clausius, Rankine and Thomson, the starting-point of all their admirable mathematical investigations.

VAPORIZATION.

140. Until, in the year 1818, Gay Lussac had indicated the contrary, it was always imagined that the temperature of boiling remains constant for a given pressure or tension. He found it to vary with the nature of the containing vessel, and to be higher in glass than in metals. He also drew attention to the irregular and explosive manner in which boiling takes place for many substances.

Subsequently, in 1842, Marcet found that, if glass vessels be first washed with sulphuric acid, they adhere more tenaciously to water, which, therefore, requires still higher temperatures in them for its ebullition. And in 1846, Donny observed that in a glass water hammer the water may be *superheated* many degrees, and until it gives off its vapour explosively.

Also, in 1863, Dufour further investigated this subject and found that when a portion of one liquid is completely enveloped by another liquid less volatile, the former may be heated far above its temperature of normal ebullition without vaporization. The smallest bubble of gas, or of escaping vapour, however, at once changes the whole phenomenon.

Another curious and interesting fact has been studied by Abel, who finds that, when the chloride of nitrogen is covered by air only, it may be exploded gently, pushing back the incumbent atmosphere without breaking a containing glass vessel; but if it be coated by a mere film of water, then it will explode with enormous and destructive violence.

In this curious fact, it would seem that the inertia of the film of water must play an important part, giving time for the explosion to extend through the entire mass; but when air alone is pushed away the manner of explosion appears to be by successive superficial layers or films, and therefore gradual and gentle.

Here we cannot fail to think of the analogy presented to these facts by the well-studied phenomena which occur in the comparative use of gunpowder and of those more violently explosive substances called bursting powders. One of the chief advantages of gunpowder in its varied uses, and absolutely necessary in artillery, being its gradual, or successive, mode of combustion, grain by grain; while bursting powders ignite in mass simultaneously, and consequently act with uncontrollable violence. The inertia effect of the film of water in the experiments of Abel also bears a striking analogy to that of the small quantity of sand used to cover the charge of powder in blasting rock.

Another phenomenon equally curious and important is that first observed by Cagniard de la Tour: that water and other liquids highly heated in strong confining vessels first expand as liquids and then at definite temperatures and pressures pass wholly into gas or vapour, leaving no surface of separation or visible liquid portion. This very remarkable change has quite recently been studied with great care by Dr. Andrews for carbonic acid. The results of whose experiments tend to show that fluids pass from the liquid into the gaseous state by continuous and imperceptible degrees, and not discontinuously, or *per saltum*, so that they are connected as it were by every intermediate state or degree of fluidity; just as some solids pass into liquids through every degree of viscidity,—butter and pitch for instance.

The experiment of Donny casts a faint light upon those horribly violent explosions of steam boilers which sometimes occur at temperatures and pressures too low apparently to account for their evidently enormous development of force; and which have not yet been sufficiently investigated.

Here too we should perhaps allude briefly to that repulsive action of overheated metals which prevents water and other liquids from enfilming or adhering to them, and thus allows their own cohesion to form small portions into drops which roll freely over the heated surface of the metal; a circumstance which led Boutigny and some others to infer very erroneously a fourth state of matter, by them called *the spheroidal state.* The spheroidal form of any liquid drop, or bubble, is, however, merely the mechanical consequence of cohesion, acting between its molecules and shaping its surface into equipotential forms of equilibrium; which will be spherical, only when there is neither rotation, nor any external force of distortion, such as gravity, or atmospheric resistance, acting upon its particles and that unequally. No one acquainted with the researches of Laplace and Poisson on capillarity can be for a moment misled by this imaginary *spheroidal state* of Boutigny.

The researches of chemists tend to prove that all solids may be by heat converted into liquids and vapours; some such as camphor, calomel, corrosive sublimate, chloral and ice passing apparently into vapour without intermediate liquefaction. With such exceptions, the familiar changes of ice into water, and of water into vapour are, therefore, typical for all bodies.

In many instances the phenomena called by chemists *allotropism* and *isomerism* are certainly thermodynamic; take for instance the effects of heat upon caoutchouc when distilled, converting it entirely into various liquids, separable by redistillation at different temperatures, but all called by the common name *caoutchoucine,* and all identical in substance and capable of being converted by still higher temperatures into olefiant gas. Or again, take the numerous other and quite similar compounds of carbon with hydrogen, presented to us in the products obtained from petroleum, gas tar, etc. ; some of which are very interesting in their applications, especially those which give the brilliant new dyes known as *analine* colours. For all of which chemical changes the purely thermal phenomena are as yet unknown.

We cannot, however, fail to see that the various definite compounds, called *caoutchoucine,* are but intermediate steps of *stable equilibrium* between viscid caoutchouc and olefiant gas; links as it were of a chain connecting the liquid and gaseous states of the same body, and which differ probably in their heat-potentials, or *latent* heat. They also may vary in the quantity and manner of their union with the luminiferous ether pervading space and all transparent substances; variations which chemists generally ignore, only because they are imponderable; but imponderability is no proof of non-existence. And as air in air, or water in water, weighs zero, so may ether in ether; even for it, therefore, weight as well as inertia may exist, though hitherto undetected. Could we but condense ether, the famous experiment of Galileo might be repeated upon it, if the extraneous forces be not in equilibrium.

As liquids pass into vapour when heated, or when, as in the art of refining sugar, evaporation in vacuo is made to take place at low temperatures by removal of pressure, so *conversely* vapours are converted by combined cold and pressure into liquids. In 1823, Faraday first liquefied chlorine and some other gases by cold and pressure in glass tubes, but subsequently he thus succeeded in liquefying all the known gases, except those few only which are called permanent; thereby proving, that there is no physical distinction between a gas and a vapour.

SATURATED VAPOUR.

141. Though such irregularities as the above-mentioned do occur, yet generally liquids pass into vapour, and gases liquefy, under normal pressures and temperatures.

Thus for any given pressure what is called boiling, or ebullition, usually takes place at a fixed temperature. But this boiling is only the mechanical agitation caused when the expansive force of the vapour becomes sufficient to overcome the external pressure, which is usually that of the atmosphere, and therefore lower on high mountains.

At any and all temperatures and pressures, from ice as well as from water and other liquids, vapour escapes in the state known as that of saturation. That is to say, of such maximum density that any increase of pressure, or decrease of temperature, will immediately cause partial condensation, while reverse changes produce additional vaporization.

When aqueous vapour, in contact with the surfaces of cold bodies, becomes chilled down below this state of saturation, it deposits upon them in the condensed form of *dew* or *frost;* the corresponding temperature is therefore called the *dew point.* And the formation of mists, clouds, rain, snow, etc., is due to the operation of like causes of condensation.

Hence, there exists for all saturated vapours, between the temperature, pressure, and specific volume, or density, a relation perfectly determinate, though unknown, which we can only express by an arbitrary function,

$$\phi\,(pvt) = 0\,;$$

but for which Regnault has given the following empirical formula, first proposed by Biot,

$$\log p = a - ba_{\prime}^{x} - ca_{\prime\prime}^{x},$$

in which

$$x = 20 + t,$$

and t denotes the temperature centigrade. For the constants Regnault gives the following values:

$$a = 6.2640348,$$
$$\log b = 0.1397743,$$
$$\log c = 0.6924351,$$
$$\log a_{,} = \bar{1}.994049292,$$
$$\log a_{,,} = \bar{1}.998343862.$$

And he also gives the following table, more convenient generally than any formula:

ATMOS.	TEMP.	ATMOS.	TEMP.	ATMOS.	TEMP.
1	100°	8	171°	15	199°
2	121	9	176	16	202
3	134	10	180	17	205
4	144	11	185	18	208
5	152	12	188	19	210
6	159	13	192	20	213
7	165	14	196	21	215

LATENT HEAT OF VAPOUR.

142. The heat which does the work of liquefaction for solids, or of vaporization for liquids, was called by Dr. Black *latent*, because it becomes *emmagazined* as energy, without causing any variation of the temperature. Thus stored up, it is capable of being liberated, or of becoming free to affect the temperature of other bodies. Hence, when steam is condensed by contact with cold bodies, it is found that it can heat up nearly five and a half times its own weight of water from 0° to 100° centigrade, or from freezing to boiling. It is, therefore, said to absorb or to emit nearly 550 units of *latent* heat. Evidently, this quantity is greater than is requisite to heat substances above redness; and we see how admirably steam is adapted to warm buildings; through which it is conveyed in iron pipes, which act as efficiently as if they were red-hot, but need never be made hotter than 100° C.; thus obviating all danger to property of destruction by fire.

It may readily be shown that the latent heat of vaporization is a *maximum* when this change takes place normally. For this, let the unit weight become vapour at the temperature t_0 and under the constant pressure p_0, and let this vapour be afterwards heated to the higher temperature t, the requisite heat will be

$$Q = \lambda + \int_{t_0}^{t} c_, dt;$$

in which λ is the latent heat of vaporization, and $c_,$ is the specific heat of the vapour.

Again, let the liquid be superheated, as in the experiment of Donny, or in those of Dufour, to the temperature t, and let it then pass into vapour. All of these changes being supposed to occur under constant pressure. Then for this second mode of change, the requisite heat is

$$Q = \lambda_, + \int_{t_0}^{t} c dt.$$

As the initial and final states are identical for both cases, it is clear that the value of Q is the same for each. Hence we have

$$\lambda = \lambda_, + \int_{t_0}^{t} (c - c_,)\, dt,$$

which proves that λ is greater than $\lambda_,$, for it is found by observation that the specific heat of liquids c exceeds $c_,$, that of their vapour.

This may be illustrated by a diagram, in which the horizontal line *ac* indicates the constant pressure, and the area *acc'a'* measuring Q is the same for both modes of change. The work due to λ is denoted by the area *abb'a'*, greater than that of $\lambda_{,}$, or the area *dcc'd'*; and the area *add'a'*,

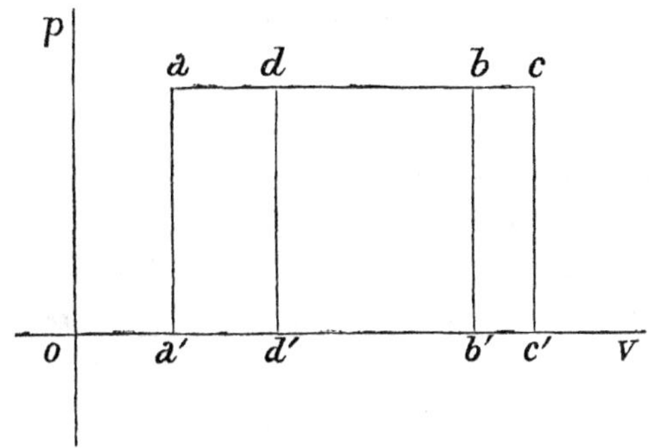

showing the superheating from t_0 to t of the liquid, is greater than *bcc'b'*, which shows the superheating of the vapour.

TOTAL HEAT OF VAPOUR.

143. We have already, in the historical introduction of this work, remarked upon the erroneous hypothesis of Sir James Watt that, for water heated from zero to any temperature t, and then converted at this temperature t into vapour, the sum of the *free* and *latent* heat due to these changes respectively is always constant. And we have also stated that this error was corrected by Regnault, who found for such a change the simple law

$$\theta = a + bt\,; \qquad (157)$$

in which t denotes the temperature, and a and b are constants, whose numerical values his observations show to be very accurately those of the formula,

$$\theta = 606.5 + 0.305t. \qquad (158)$$

The hypothesis of Watt would make θ equal to the constant a only.

To the heat indicated by formulas (157, 158) and requisite first to heat a unit of water from 0° to t° centigrade and then to evaporate it at t°, Regnault has given the fictitious name of *the total heat* of vaporization; which we here give only because, though false, it is much used by some writers, and needs, therefore, careful definition to guard against errors which are apt to flow from the false use of common words.

To obtain from these formulas of Regnault one for the latent heat of vaporization of water at the temperature t°, it is evidently only necessary to subtract from θ the heat required to raise the unit of water from 0° to t° centigrade; which gives

$$\lambda = \theta - \int_{t_0}^{t} c dt. \tag{159}$$

Although c, the specific heat of water, is equal to unity only between 0° and 1° centigrade, yet it varies so slightly that for all practical and many theoretical purposes it may be regarded as unity and constant. Hence we obtain,

$$\lambda = 606.5 - 0.695t; \tag{160}$$

as the numerical formula given by the data of Regnault for computing the heat requisite to evaporate the unit weight of water at any given temperature t centigrade.

For very exact purposes, Regnault gives for the term of reduction

$$\int_0^t c dt = t + 0.00002t^2 + 0.000000t^3. \tag{161}$$

In applying formula (160) it is sometimes necessary to separate the *internal* from the *external* work. For this

$$\lambda = A\left(\Delta U + \int_0^t p dv\right),$$

and if we put p equal to the mean atmospheric pressure, 760^{mm}, the data of Regnault give for water,

$$A\int_0^t pdv = 31.10 + 0.096t;$$

from which

$$A\Delta U = 575.40 - 0.791t. \tag{162}$$

The greater part of the latent heat of evaporation is therefore consumed in overcoming cohesion; and only about 18 per cent thereof is available for external work.

CHAPTER X.

ON STEAM AND OTHER VAPOURS.

Having given the principal physical phenomena of vaporization, we now propose to obtain and discuss the general formulas needed for their practical and theoretical applications; and for which we are indebted chiefly to Rankine, Thomson and Clausius.

FORMULAS FOR ELEMENTARY VAPORIZATION.

144. In the boiler of any steam engine, after all air has been driven out, there exists only a mass of water and of saturated steam. For such a variable mixture of liquid and vapour any addition of heat will cause a variation of pressure and temperature, both in the vapour and in the liquid, and these changes will be denoted for an unit of weight by the general formula

$$c d\tau + h dp.$$

But as the pressure is a function of the temperature this will become for the vapour, or steam,

$$m d\tau = \left(c + h\frac{dp}{dt}\right)d\tau,$$

or

$$m = \left(c + h\frac{dp}{dt}\right). \qquad (162)$$

An important formula for the coefficient of temperature m, which is called the *specific heat of saturated vapour.*

Analogously, we have for an unit weight of the liquid the precisely similar formula

$$m' = \left(c' + h'\frac{dp}{dt}\right). \tag{163}$$

Let now an unit of the mixture composed of a variable part x of saturated vapour, and $1 - x$ of the corresponding liquid, pass from the physical condition denoted by the independent variables x and t to that for which they will become $x + dx$ and $t + d\tau$. Then if λ be the latent heat, or coefficient of vaporization,

$$dQ = \lambda dx + mxd\tau + m'(1 - x)d\tau \tag{164}$$

will evidently be the heat required.

This equation is the same as

$$\frac{dQ}{\tau} = \frac{\lambda}{\tau}dx + [m' + (m - m')x]\frac{d\tau}{\tau}. \tag{165}$$

But

$$\frac{dQ}{\tau} = d\phi$$

is an exact differential; and therefore,

$$\frac{m - m'}{\tau} = \frac{d\left(\frac{\lambda}{\tau}\right)}{d\tau}. \tag{166}$$

Which, by differentiation, gives

$$\frac{d\lambda}{d\tau} + m' - m = \frac{\lambda}{\tau}; \tag{167}$$

an important formula of Sir W. Thomson between the coefficients λ, m, m', from which x has been eliminated.

But the formula of Clapeyron (156) gives

$$\frac{\lambda}{\tau} = A(v - v_0)\frac{dp}{d\tau};$$

and therefore

$$\frac{d\lambda}{d\tau} + m' - m = A\,(v - v_0)\,\frac{dp}{d\tau}. \tag{168}$$

This equation (168) first given by Clausius, expresses in the most general and complete manner the relations between the latent heat of vaporization and the changes of pressure, volume and temperature.

145. The equations of Thomson (165, 166) give

$$\frac{dQ}{\tau} = m'\frac{d\tau}{\tau} + x d\left(\frac{\lambda}{\tau}\right) + \frac{\lambda}{\tau}\,dx\,;$$

or

$$d\phi = m'\,\frac{d\tau}{\tau} + d\left(\frac{\lambda x}{\tau}\right). \tag{169}$$

From which as m is a function of the temperature only, we obtain

$$\phi = \frac{\lambda}{\tau}\,x + \int_{\tau_0}^{\tau} m'\,\frac{d\tau}{\tau}\,; \tag{169'}$$

and if ϕ be constant, this will be the equation of an adiabatic line.

146. If we denote by s the specific volume of the saturated vapour, and by s_0 that of the liquid, then for the unit mass of their mixture, the total volume will be

$$v = sx + s_0\,(1 - x)\,,$$

or

$$v = s_0 + (s - s_0)\,x\,; \tag{170}$$

for which x and τ are the independent variables. Hence

$$dv = \frac{ds_0}{dt}\,d\tau + x\,\frac{d\,(s - s_0)}{dt}\,d\tau + (s - s_0)\,dx.$$

And if we combine this equation with (165) and with the general formula

$$dQ = AdU + Apdv,$$

we shall readily obtain for internal energy,

$$AdU = [m' + (m - m')\, x]\, d\tau$$
$$- Ap\left[\frac{ds_0}{dt} + x\frac{d\,(s - s_0)}{dt}\right] d\tau$$
$$+ [\lambda - Ap\,(s - s_0)]\, dx.$$

But as t and x are independent variables, this equation gives the partial derivatives

$$A\frac{dU}{dt} = m' + (m - m')\, x - Ap\left[\frac{ds_0}{dt} + x\frac{d\,(s - s_0)}{dt}\right],$$

$$A\frac{dU}{dx} = \lambda - Ap\,(s - s_0).$$

Whence

$$A\frac{d^2U}{dtdx} = m - m' - Ap\frac{d\,(s - s_0)}{dt};$$

$$A\frac{d^2U}{dxdt} = \frac{d\lambda}{d\tau} - A\,(s - s_0)\frac{dp}{dt} - Ap\frac{d\,(s - s_0)}{dt}.$$

And equating these values we again find

$$\frac{d\lambda}{d\tau} + m' - m = A\,(s - s_0)\frac{dp}{dt}. \qquad (171)$$

Which is evidently identical with equation (168), for s and s_0 are the same as v and v_0; but this more analytical demonstration, due to Clausius, gives another and elegant mode of arriving at the same result.

And if we combine the second member of the equation just obtained with the second of equation (169), we have

$$\lambda = A\tau (s - s_0) \frac{dp}{dt} ; \qquad (172)$$

or the equation of Clapeyron, changed from the form (156) in letters only.

GENERAL APPLICABILITY OF FORMULAS.

147. Though we shall pass in detailed review many of the more important applications and consequences of the general formulas just deduced, yet to give the reader a clear idea of their signification, the following remarks may not here be inappropriate.

It is evident that if, instead of the vaporization of a liquid, we assume the phenomenon to be that of the liquefaction of a solid, ice for instance melting into water, then all the above formulas applicable in the one case become equally so in the other.

Equation (172) gives

$$dt = A \frac{\tau}{\lambda} (s - s_0)\, dp, \qquad (173)$$

which shows that, if s be greater than s_0, as for the vaporization of water, and for the liquefaction of some solids, then increase of pressure will cause an elevation of the boiling or melting point; but when s is less than s_0, which happens for melting ice, then increase of pressure will lower the temperature of the melting point.

Prior to the deduction of this consequence from the mechanical theory of heat, it was imagined and believed that the melting points of solids do not depend at all upon pressure. And, though it was known that water is often chilled below its freezing point without solidification, yet ice was believed to melt always at a fixed temperature; which was, therefore, adopted as one of the fixed points for the scales of ordinary thermometers. That melting, as well as boiling, varies with pressure, was therefore a new and important discovery, first deduced theoretically, or predicted, by Prof. James Thomson in 1848; whose distinguished brother, Sir W. Thomson, soon afterwards verified that prediction by experiment. Thus do true physical theories anticipate observation, and prove in the hands of able mathematicians powerful means of valuable and unexpected discovery.

148. Another and even a more interesting and valuable discovery, or prediction, is that made, in 1850, simultaneously by Rankine and by Clausius, as a mathematical deduction from theory, and which, in 1853, Hirn verified experimentally; to wit, that saturated steam expanding in the cylinder of an engine loses latent heat converted into mechanical work, and consequently becomes partially liquefied. To this discovery we have already referred in the historical introduction to this treatise; but it is of such importance that it demands complete discussion; nor can that be done better elsewhere than here.

LIQUEFACTION OF EXPANDING SATURATED VAPOUR.

149. The data of Regnault for the latent heat of steam prove that in equation (167) the second member is greater than

$$\left(\frac{d\lambda}{d\tau} + m'\right);$$

hence in the first member m, the specific heat of saturated steam, must be negative.

We will give the interpretation of this *negative* value of m in the very words used by Rankine (Trans. Roy. Soc. Edin., Feb. 1850, t. xx., p. 171) to announce it to the scientific world, words which recorded it forever:

"The kind of specific heat under consideration is a *negative* quantity; that is to say, that, if a given weight of vapour at saturation is increased in temperature, and at the same time maintained by compression at the maximum elasticity, the heat generated by the compression is greater than that which is required to produce the elevation of temperature, and a surplus of heat is given out; and on the other hand, if vapour at saturation is allowed to expand and at the same time maintained at the temperature of saturation, the heat which disappears in producing the expansion is greater than that set free by the fall of temperature, and the deficiency must be supplied from without, *otherwise a portion of the vapour will be liquefied, in order to supply the heat necessary for the expansion of the rest.*" (The italics here given were used by Rankine.)

"This circumstance is obviously of great importance in meteorology and in the theory of the steam-engine. There is as yet no experimental proof of it." (Since so proved, in 1853, by Hirn.)

"It is true that in the working of non-condensing engines, it has been found that the steam which escapes is always at the temperature of saturation corresponding to its pressure, and carries along with it a portion of water in the liquid state; but it is impossible to distinguish between the water which has been liquefied by the expansion of the steam, and that which has been carried over mechanically from the boiler."

We have already stated how Hirn, in 1853, by using a hollow cylinder, connected with a boiler and with the air by tubes and stopcocks, and closed at its ends by glass plates, was enabled to see the condensed clouds which form when saturated steam expands; thus verifying Rankine's admirable theoretical conclusions, and thereby completing the most important discovery concerning the steam-engine, which has been made since the day of Dr. Black and of Sir James Watt, its grand inventor.

Well may the glorious old University of Glasgow now be proud that, as in 1750, from those eminent men within her walls, the world received its knowledge of latent heat and of the steam-engine in its present form, so likewise, a century later, in 1850, to her distinguished professors, Rankine and Thomson, has that world been also indebted for much of what has been achieved in perfection of that knowledge.

150. The discovery of Rankine, and the importance of the *negative* value of the coefficient m in the theory of steam, may be rendered clearer by the following analytical discussion thereof.

Let any quantity of saturated vapour suffer the change $d\tau$, and denote by dq the heat due to such a variation of temperature in a unit of mass. Then will

$$dq = m d\tau.$$

Take as independent variables the heat q and the specific volume v, or the reciprocal of the density; we have

$$dq = m \frac{dt}{dv} dv,$$

which gives

$$\frac{dv}{dt} = m \frac{dv}{dq}.$$

Observation proves that the first member of this equation is negative, for the density of saturated steam increases, or its specific volume decreases, with the temperature. And as m is also negative for steam, or aqueous vapour, it follows that the ratio of dv to dq must always be positive; or they must both have the same algebraic sign.

Hence, when compression occurs, dv is negative and dq will also be *negative,* or heat will be emitted, causing the vapour to become *superheated.*

But if expansion take place, dv and dq will both be positive, or heat *must be absorbed.* If, therefore, as in the cylinder of an engine, this expansion be so rapid that time is not given to allow heat to be absorbed from other bodies, then one portion of the saturated steam will be liquefied to furnish latent heat to the remainder and preserve this in the state of saturated vapour.

SPECIFIC HEAT OF SATURATED VAPOUR.

151. As the specific volume v is a function of the temperature t, the increment of heat given to a unit of any saturated vapour is

$$dq = mdt = \left(\frac{dq}{dt} + \frac{dq}{dv} \cdot \frac{dv}{dt}\right) dt.$$

Therefore, the coefficient m, or quantity within the brackets, is a binomial of which the first term is positive; but its second term is composed of two factors, of which one is positive and the other negative. Hence the two terms of the binomial are of opposite signs, and its value may be either positive or negative for different kinds of vapour.

The observations of Regnault and others upon the specific heat m' of different substances and upon the latent heat λ of their vaporization, enable us to calculate for them respectively the values of m, the coefficient of dt. This is readily done by aid of the equation (167), put for this purpose under the form

$$m = \left(\frac{d\lambda}{d\tau} + m'\right) - \frac{\lambda}{\tau}; \tag{174}$$

in which the term within brackets is simply the derivative of what Regnault has called *the total heat of vaporization.*

In this manner Clausius first obtained for steam the following table:

CENTIGRADE. $t°$.	COEFFICIENT. m.
0°	— 1.916
50°	— 1.465
100°	— 1.133
150°	— 0.879
200°	— 0.676

And similarly, from observations by Regnault, others have since calculated the values of m for the different liquids of the following table:

$t°$.	m.
ETHER.	
0°	+ 0.116
40°	+ 0.120
80°	+ 0.128
120°	+ 0.133
SULPHIDE OF CARBON.	
0°	— 0.184
40°	— 0.171
80°	— 0.164
120°	— 0.163
CHLOROFORM.	
0°	— 0.107
40°	— 0.047
80°	+ 0.001
120°	+ 0.050
160°	+ 0.072
BENZINE.	
0°	— 0.155
70°	— 0.038
140°	+ 0.048
210°	+ 0.115

These results show that benzine and chloroform behave like sulphuric ether at high temperatures, but like water and the sulphide of carbon for low temperatures. Also we see that there must exist for each of them a particular temperature, called that of *inversion*, where the sign of m changes from negative to positive.

This temperature of *inversion* is readily calculated from equation (174), by making in it m equal to zero; and is for benzine about 120° centigrade.

It is also easy to see that all these numbers *increase* with the temperature; and therefore, that for water there should exist a temperature of inversion which Rankine calculates at 520° cent.; and for ether it would be — 116° centigrade nearly.

At this temperature of inversion, for which m is zero, a slight compression or expansion will not cause either superheating or condensation. But below it expansion produces liquefaction; and compression causes saturated vapour to become superheated. While precisely reverse phenomena take place at temperatures above this point of inversion.

152. The conclusion just stated, that a vapour for which m is positive behaves under compression, or expansion, in a manner opposite to that of one for which m is negative, was simply and beautifully verified by Hirn. For ether m is positive, and for the sulphide of carbon and for water it is negative.

Hirn therefore took a glass globe into which he put some ether; and then attached to it a pump, or syringe, of sufficient size. This he then heated by plunging it entirely under hot water; and when all air had been driven out by the vapour of the ether, through a stopcock for that purpose, the piston of the syringe, which had been pushed outwards by the vapour, was quickly forced inwards; and immediately a cloud of condensed vapour was observed in the glass globe. When this same experiment was tried with the vapour of sulphide of carbon, compression produced no cloud whatever, but the vapour in the globe retained, as it should do, its perfect clearness and transparency.

We have already given (see article 10) calculations which show of what great practical importance it was to have discovered the liquefaction of saturated steam by expansion in the working cylinders of engines. And it is, therefore, unnecessary to recur to them here.

DENSITY OF SATURATED STEAM.

153. From the equation of Clapeyron, or formula (172), it is clear that the specific volume s, or its reciprocal the density, may be readily calculated from Regnault's data for the latent heat λ and for p the tension or pressure.

The experiments made by MM. Fairbairn and Tate, to which we have referred in article (83), allow such calculations to be compared with the results of observation; and this has been done by Clausius, from whom we take the following table, in which the values of s express in cubic metres the volume of saturated vapour produced by a kilogramme of water:

TEMP. $t°$ C.	VALUES OF s.		
	Calculated.	*Observed.*	*Law of Mariotte.*
58°.21	8.23	8.27	8.38
77.18	3.74	3.72	3.84
92.66	2.11	2.15	2.18
117.17	0.947	0.941	0.991
130.67	0.639	0.634	0.674
144.74	0.437	0.432	0.466

The apparatus used by MM. Fairbairn and Tate was necessarily so complicated that we shall not attempt any description of it.

And when we bear in mind how difficult it is to observe values of s with any accuracy, it will be seen that the concordance between the observed and the calculated values of s in the above table is quite satisfactory, but that they both differ much from values given for s by the law of Mariotte previously and erroneously adopted.

As the values of s decrease with the temperature, it is evident that the densities, their reciprocals, must increase, and that they follow a very different law from that of Mariotte.

The values of s being the number of cubic metres of vapour which weigh one kilogramme, it is easy to get from them the weights of one cubic metre. And then, by comparing these with the weight of a cubic metre of air, under like conditions of temperature and pressure, we may obtain the ratios of their relative densities. Formerly, the ratio of the relative density of steam to that of air was falsely supposed to be constant and equal to 0.622; but this error was due to the ignorant assumption that they both obey the law of Mariotte.

EMPIRICAL FORMULA OF ZEUNER.

154. It is of course very desirable to obtain for saturated steam a formula expressing the relation between the pressure, specific volume and temperature, or in other words to ascertain the form of the function

$$\phi(pvt) = 0.$$

This has been done to a certain degree of approximation by Zeuner; who has given for that purpose the empirical formula

$$ps^n = 1.704; \tag{175}$$

in which

$$n = 1.0646;$$

and as both the pressure p and the specific volume s are functions of the temperature it can be omitted.

To show to what degree this formula may be relied upon, Zeuner has computed for the densities ω, or reciprocals of s, the values of the following table, as given by his formula and by observation:

DENSITY OF STEAM.					
p	ω *Calculated.*	ω *Observed.*	p	ω *Calculated.*	ω *Observed.*
1	0.606	0.606	6	3.262	3.263
2	1.162	1.103	8	4.274	4.274
3	1.701	1.703	10	5.270	5.270
4	2.229	2.230	12	6.255	6.254

The pressures p are here expressed in atmospheres.

It is evident from this table, that formula (175) may be regarded as an approximation sufficiently exact for all practical purposes. But as it is entirely empirical, it is far better for exact theoretical purposes to use for calculating s the equation (172) of Clapeyron; which we know to be absolutely true and applicable to bodies in every physical state.

GEOMETRIC CURVE FOR SATURATED VAPOUR.

155. With the values of s given by the law (172) of Clapeyron, or by formula (175), we readily construct the curve whose consecutive points represent corresponding physical states of an unit of saturated vapour, for which the pressure, density, and temperature so vary that it remains dry and saturated, without liquefaction or superheating. And formula (175) shows that this curve of saturation must be a species of hyperbola referred to asymptotic axes p and v, and which varies more rapidly for ordinates p than for abscissas v, because the constant exponent n exceeds unity.

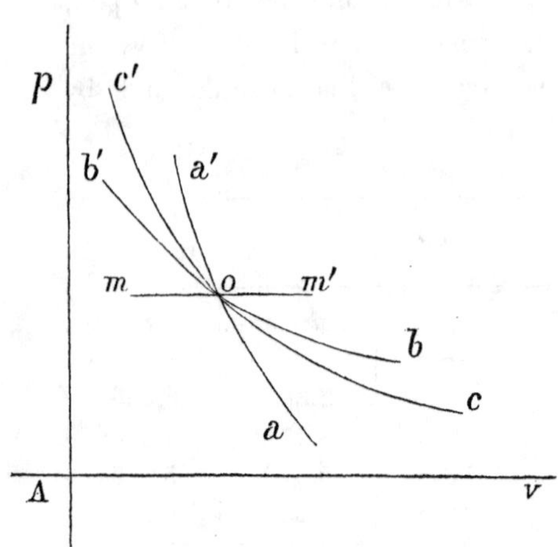

Let now cc' represent this curve of saturation, and let it be cut at any point o by the horizontal line mm' and by the adiabatic lines aa' and bb'; the former aa' for steam, and the latter bb' for the vapour of sulphuric ether.

When a quantity of saturated vapour is superheated under constant pressure, both its temperature and its volume increase; and such a change may always be denoted by the line om'. Reversely, when it is chilled a portion liquefies and the volume diminishes, which may be represented by the line om. We see, therefore, that the curve cc' separates the angular space between the axes into two regions, one to the right and above the curve indicating *superheated* vapour, the other to the left and below the curve showing liquefaction.

Hence of the two adiabatic lines cutting cc' dissimilarly at o, one aa' shows for steam the liquefaction caused by sudden expansion and the superheating due to compression; while the other bb' indicates for ether phenomena precisely reverse.

Here it may be not inappropriate to refer to the cloud of condensed vapour which forms whenever a jet of steam escapes into the air, as a familiar phenomenon due to a sudden, and therefore to an adiabatic, change of volume, one portion of steam giving up its latent heat required for the expansion of another and thus becoming liquefied, rather than to any chilling effect caused by contact with cold air, which is well known to be an almost perfect non-conductor.

To those acquainted with the thermodynamic theory of storms, urged by the late Mr. Espy,—a theory to which, in the opinion of the writer, proper attention and respect was not paid,—the adiabatic formation of clouds, rain, hail, and snow in the upper atmosphere will appear as a rich field of meteorological research, of which Rankine caught only a faint glimpse; one distinct enough, however, to cause him, in one of the passages above quoted, to cite the explanation of such meteorological phenomena as an important application of his discovery.

ADIABATIC CHANGES IN SATURATED STEAM.

156. We have obtained for saturated vapour, or steam, the fundamental equations (168) and (172), which in the mechanical theory of heat as applied to the steam-engine take the place of the two general laws of Joule and Carnot; and we now propose to determine from these equations the variable quantity of vapour x, the volume v, and the work S, considered as functions of the temperature when, as in either end of the working cylinder of an engine closed by its moveable piston, the variable mass, partly liquid and partly vapour, changes its volume adiabatically, without loss or gain of heat, but performing external work positively or negatively. And in these demonstrations we shall follow chiefly Clausius ("Théor. Méc. de la Chaleur," t. i, p. 180, 2d edit., Paris, 1868), who first gave them.

157. PROBLEM I.—*To determine x, the quantity of vapour, for any temperature τ, when x_0, the quantity for a given temperature τ_0, is known.*

It is perhaps scarcely necessary now to say that the letter τ in the notation we adopt always denotes the absolute temperature corresponding to that denoted by t for the ordinary centigrade scale; or that λ is the latent heat of steam.

As the change is adiabatic, equation (164) will become

$$\lambda\, dx + x\,(m - m')\, d\tau + m' d\tau = 0;$$

but, by equation (167),

$$(m - m') = \frac{d\lambda}{d\tau} - \frac{\lambda}{\tau};$$

hence, by substitution,

$$(\lambda dx + x d\lambda) - x\frac{\lambda}{\tau}\, d\tau + m' d\tau = 0;$$

or

$$d\,(\lambda x) - x\frac{\lambda}{\tau}\, d\tau + m' d\tau = 0. \qquad (176)$$

Dividing this equation by τ, and observing that

$$\frac{d\,(\lambda x)}{\tau} - \frac{x\lambda}{\tau^2}\, d\tau = d\left(\frac{x\lambda}{\tau}\right),$$

we obtain

$$d\left(\frac{x\lambda}{\tau}\right) + m'\frac{d\tau}{\tau} = 0.$$

Whence, by definite integration,

$$\frac{\lambda}{\tau}\, x = \frac{\lambda_0}{\tau_0}\, x_0 - \int_{\tau_0}^{\tau} m' \frac{d\tau}{\tau}. \qquad (177)$$

As the specific heat of a liquid m' varies very slowly with the temperature, it may generally be considered constant, which gives

$$\frac{\lambda}{\tau}\, x = \frac{\lambda_0}{\tau_0}\, x_0 - m' \log \frac{\tau}{\tau_0}. \qquad (178)$$

From which equation the data of Regnault allow x to be easily calculated. As an example, Clausius gives the values of x of the following table, computed for a quantity of steam in a cylinder, saturated and without liquid at 150° C., but becoming superheated when compressed, and partially liquefied by dilatation:

$t^\circ =$	150°	125°	100°	75°	50°	25°
$x =$	1	0.956	0.911	0.866	0.821	0.776

158. Simple as is the above demonstration of Clausius for equation (177), a yet simpler one flows from the equation (169) of Thomson, which for an adiabatic change gives

$$m' \frac{d\tau}{\tau} + d\left(\frac{\lambda x}{\tau}\right) = 0;$$

whence, immediately, by integration,

$$\frac{\lambda}{\tau} x = \frac{\lambda_0}{\tau_0} x_0 - \int_{\tau_0}^{\tau} m' \frac{d\tau}{\tau},$$

as before.

159. Problem II.—*To find the change of volume.* For this we denote by s_0, as in equation (170), the volume of a unit of the liquid in contact with its saturated vapour, and by s the volume of a unit of that vapour. Then we have, as above,

$$v = s_0 + (s - s_0)\, x;$$

in which s_0 may be considered constant, for comparatively with s its variation with temperature is exceedingly small.

Putting now, for brevity,

$$u = s - s_0, \qquad (179)$$

it is clear that we have only to find the product ux. For this, substitute in equation (178) the value given for λ in equation (172), and we get:

$$xu\frac{dp}{dt} = x_0u_0\left(\frac{dp}{dt}\right)_0 - \frac{m'}{A}\log\frac{\tau}{\tau_0}; \tag{180}$$

from which equation, and the data of Regnault for the tension p of saturated steam, we readily obtain the value of ux, to which s_0 must be added to find the volume. In this manner Clausius computed for v the values given in the table below (see Art. 160). And to compare them he also computed the analogous values v' given by the false hypothesis previously assumed, that steam which expands remains saturated without partial liquefaction and obeys the laws of Charles and Mariotte.

160. Problem III.—*To obtain the external work done by an adiabatic change.* Its value will be

$$S = \int_{\tau_0}^{\tau} pdv.$$

But by equation (179), we have

$$dv = ds_0 + d\,(xu),$$

and s_0 may be considered constant; therefore

$$pdv = d\,(xup) - xu\frac{dp}{dt}\,d\tau,$$

as p is a function of t. But by the equation of Clapeyron (172), we have

$$u\frac{dp}{dt} = \frac{1}{A}\cdot\frac{\lambda}{\tau};$$

whence

$$pdv = d\,(xup) - Ex\frac{\lambda}{\tau}d\tau.$$

And, therefore, from equation (176), we have

$$pdv = d\,(xup) - E\,[d\,(\lambda x) + m'd\tau].$$

From this equation, by definite integration, we find the work

$$S = xup - x_0u_0p_0 - E\,[\lambda x - \lambda_0 x_0 + m'\,(\tau - \tau_0)]\,; \quad (181)$$

an expression in which xu and λx are known by the equations already found for them, and from which the values of S are therefore readily calculated. If for perfectly exact results the hypothesis that s_0 is constant be rejected, then we must add to the values given for S by the equation just found those of the integral

$$\int_{\tau_0}^{\tau} pds_0.$$

From the equation (181) and those preceding, Clausius calculated the results given in the following table:

$t°$	x	v	v'	S
125°	0.956	1.88	1.93	11,300
100	0.911	3.90	4.16	23,200
75	0.866	9.23	10.11	35,900
50	0.821	25.7	29.7	49,300
25	0.776	88.7	107.1	63,900

For all of the data of this table the unit of volume ($v = 1$) is that of a kilogramme, or unit of vapour ($x = 1$), at the temperature 150° C.; and the work S is given in French units of work, or in kilogrammetres.

To show the great importance of *expansion* in the work of engines, Clausius states the fact, which may be compared with the numerical values of S in the last vertical column of this table, that the work done against external pressure is 18,700 units, when a kilogramme of water is converted into steam at 150° C. and under the corresponding pressure. And it will be observed that, between the temperatures 150° and 50°, the volume becomes nearly twenty-six times its original value.

ENERGY OF A VARIABLE MIXTURE OF LIQUID AND VAPOUR.

161. As the variation of the internal energy of any mass, or system of masses, has been shown to be equal to the external work done, it is well to obtain an analytical expression for it in the case under consideration, or that of a unit mixture of a liquid in contact with its saturated vapour.

The heat required to change the temperature of the unit mass while liquid from τ_0 to τ is

$$\int_{\tau_0}^{\tau} m' d\tau.$$

Our general formula gives also for this same change

$$A\left(U_1 - U_0 + \int_{\tau_0}^{\tau} p ds_0\right);$$

whence

$$A\,(U_1 - U_0) = \int_{\tau_0}^{\tau} m' d\tau - A\int_{\tau_0}^{\tau} p ds_0.$$

Let now the fraction x pass at the temperature τ and pressure p into vapour of maximum density; the heat required will be

$$\lambda x = A\,[U - U_1 + x\,(s - s_0)\,p].$$

From which equation we get

$$A\,(U - U_0) = \lambda x + \int_{\tau_0}^{\tau} m' d\tau - Axup - A\int_{\tau_0}^{\tau} p ds_0;$$

an expression easily seen to be identical with the formula (181) of Clausius for s the external work, except in its sign, *which differs because work is negative energy.*

The formula just obtained may be rendered more convenient for application. Observing that

$$v = s_0 + xu,$$

and

$$\int p ds_0 = ps_0 - \int s_0 dp,$$

and making these substitutions we easily obtain

$$U - U_0 = E\left(\lambda x + \int_{\tau_0}^{\tau} m'dt\right) - pv + p_0 s_0 + \int_{p_0}^{p} s_0 dp. \quad (182)$$

In this equation U_0 is the initial energy, and therefore a determinate or constant quantity, though unknown.

For the definite integral, or value of U between the physical states (1) and (2), for which the temperatures are τ_1 and τ_2 the equation just found gives

$$\begin{aligned} U_2 - U_1 = E\,(\lambda_2 x_2 - \lambda_1 x_1) - p_2 v_2 + p_1 v_1 \\ + E\int_{\tau_1}^{\tau_2} m'dt + \int_{p_1}^{p_2} s_0 dp \end{aligned} \quad (183)$$

It is customary in using these formulas to simplify them by putting m' equal to c the specific heat of the liquid for constant pressure and regarding c as constant; both of which hypotheses are nearly true. Also the volume of the liquid s_0 may be considered constant. Making which changes, we have

$$U - U_0 = E\,[c(\tau - \tau_0) + \lambda x] - p\,(v - s_0) \quad (184)$$

and

$$\begin{aligned} U_2 - U_1 = E\,[c\,(\tau_2 - \tau_1) + \lambda_2 x_2 - \lambda_1 x_1] \\ - p_2 v_2 + p_1 v_1 + (p_2 - p_1)\, s_0. \end{aligned} \quad (185)$$

SUPERHEATED STEAM.

162. The experiments of Faraday having proved that all the known gases, except only those few which are called *permanent*, may readily be liquefied by cold and pressure, it follows at once that they are all only superheated vapours.

Hence, all that we have said about the thermal properties of airs, or gases, in the beginning of this work, may be generally considered applicable to superheated vapours. In fact, the solid, liquid and gaseous states do, as we have seen, pass gradually and continuously into each other. And the study of the physical properties of vapours, except near to and at their points of saturation when they are becoming liquid, is therefore only that of the laws of such gases as do actually exist. The use of an ideal or *perfect gas* as a *limit* has been sufficiently explained.

In article (83) we have mentioned the experiments of Messrs. Fairbairn and Tate, and those of Hirn, upon the density and expansion of saturated and *superheated* steam. The experiments of Regnault have also given us, for various gases and vapours, their coefficients of dilatation under constant pressure and constant volume, and their specific heat under constant pressure. And to these data we may of course apply the general formulas which we have proved for bodies in all physical states whatever.

But when we attempt to go further, and deduce from observations the form of the thermodynamic function

$$\phi\,(pvt) = 0,$$

even for ordinary gases, of which our knowledge is certainly far more perfect than it is for superheated steam, it seems impossible yet to solve that difficult problem; except approximatively and for the few gases only which approach closely to the nature of what is called a *perfect gas*, and therefore obey very nearly the laws of Charles and Mariotte; or algebraically, which follow the law

$$\frac{pv}{p_0 v_0} = \frac{\tau}{\tau_0}.$$

Both Hirn and Zeuner have, however, attempted to give formulas for superheated steam which in the present imperfect state of experimental knowledge, may be regarded as just approximations, as was formerly the law of Mariotte; which even yet is often, but improperly, employed in technical calculations.

163. Hirn simply assumes that an *isodynamic* curve, as it is called by Cazin, or one of *constant internal energy*, is an equilateral hyperbola, which will not be isothermal. And we have seen (Art. 110) that an *isothermal* line, or curve of constant temperature, will not be *isodynamic*, unless U the internal energy is a function of the temperature only; and that this would be true for a perfect gas obeying the law of Charles and Mariotte,

$$pv = R\tau,$$

is evident. But the experiments of Thomson and Joule (see 87) prove conclusively, as do also those of Regnault, that not even hydrogen obeys that law exactly. It is not possible, therefore, that for any known substance an isothermal line can obey the law of Mariotte; for which the algebraic expression is

$$pv = c,$$

and its geometric construction is an equilateral hyperbola. The hypothesis of Hirn is, therefore, simply that the product c is constant for *isodynamic* curves. The simplicity of this hypothesis is certainly a great recommendation for numerical applications; and as an approximation, it is quite close enough, except when steam or vapour approaches its condition of saturation. Then the errors of the calculated results, as compared with those of observation, become too great.

164. To determine by steam the form of a curve, which shall be either adiabatic or isodynamic, but not isothermal; let p and v be the independent variables, and put

$$Xdv + Ydp = 0;$$

whence

$$Xp\left(\frac{dp}{p} + \frac{Yv}{Xp} \cdot \frac{dv}{v}\right) = 0.$$

Or assuming

$$\frac{Yv}{Xp} = n, \tag{186}$$

we have

$$\frac{dp}{p} + n\frac{dv}{v} = 0; \tag{187}$$

which, if we make n constant, gives by integration

$$pv^n = c. \tag{188}$$

If now in

$$dQ = AdU + Apdv,$$

or in its equivalent,

$$dQ = A\left(\frac{dU}{dv} + p\right)dv + A\frac{dU}{dp}dp,$$

we suppose Q to be constant, then we shall have, as above,

$$Xdv + Ydp = 0;$$

giving by integration equation (188) as that of an *adiabatic* curve.

But if, instead of Q, we suppose the internal energy U to be constant, then our general equation evidently divides into the two following;

$$dQ = Apdv,$$

and

$$\frac{dU}{dv}dv + \frac{dU}{dp}dp = 0;$$

the latter of which, identical with

$$Xdv + Ydp = 0,$$

gives by integration equation (188), as that of an *isodynamic* curve when n is constant. While the first equation shows that, in any isodynamic change, heat absorbed is entirely converted into external mechanical work.

From the above, it is clear that the integration which gives equation (188) as the general form for an *adiabatic* curve, if Q be constant, or for an *isodynamic* curve if U be constant, depends entirely upon the hypothesis that the factor n of formula (186) is constant. An hypothesis which we have no right to assume without demonstration.

165. Rankine was the first to use for adiabatic lines the formula (188); and he seems to have been led to it by noticing that in diagrams of energy of steam-engines, drawn by the indicator of Watt, the adiabatic curves are in form *hyperbolic;* and that it is, therefore, very convenient to discuss adiabatic changes graphically by hyperbolas which are not equilateral.

Or in his own words,—"it has been deduced by trial, that for such pressures as usually occur in the working of engines, the relation between the co-ordinates is approximatively expressed by the following statement: *the pressure varies nearly as the reciprocal of the nth power of the space occupied.* The convenience of this method arises from the fact that the curve approximates to one of the *hyperbolic class;* that is, a curve in which the ordinate is inversely proportional to some power of the abscissa, as is expressed by the equation

$$pv^n = c.$$

The index n is different according to the circumstances of the case, and is to be found by trial. When n is equal to unity, the curve is an equilateral hyperbola. But in the cases which occur in the working of saturated steam n is fractional and greater than unity."

Rankine gives for n, in the adiabatic curve of saturated steam, the erroneous value 1.1111, or ten ninths. (See Rankine on Steam-Engine, p. 385, edit. 3d, 1866.) But Zeuner, who adopts Rankine's formula (188), has corrected this error, and gives for that curve $n = \frac{4}{3} = 1.3333$. Which result MM. Cazin and Hirn confirm, as agreeing very closely with their recent observations.

166. Zeuner, generalizing the hypothesis of Rankine, assumes that for all vapours and gases, the thermodynamic function

$$\phi\,(pvt) = 0$$

may, if adiabatic or isodynamic, be put under the form (188); which for the curve of saturation of steam, (Art. 154), has for the index n the value 1.0646, and for the constant c the value 1.704. And between these maximum density values for n and c in saturated steam, and those of the law of Mariotte, when n becomes a maximum and equal to unity, Zeuner supposes n to be variable and to decrease as the steam becomes more and more superheated. Geometrically, this amounts to the assumption that

$$\phi = pv^n = c$$

is an hyperbola with a variable index n, and whose variables p, v, and n are all functions of the temperature. The curve also becomes equilateral for the limit value $n = 1$ required by Mariotte's law.

Evidently, however, formula (188) must be considered as empirical only, whenever the value of the factor (186) is for convenient integration assumed without demonstration to be constant. And the only apology that can be made for the adoption of this formula (188) is that of Rankine, to wit its practical utility and convenience.

The general equation of an isodynamic curve

$$Xdv + Ydp = 0,$$

or its equivalent,

$$dU = \frac{dU}{dv}dv + \frac{dU}{dp}dp = 0,$$

is evidently an exact differential and therefore integrable. But to integrate it we must know X and Y in terms of v and p, and these variables must be separated. And practically this is impossible until the function U is given by observation.

167. Zeuner and Hirn have more recently both given another approximative equation for superheated steam of the form

$$pv = B\tau - Cp^a,$$

which Zeuner finds to agree very accurately with observed data, and which Hirn endeavors to base upon theoretical reasoning. (See Hirn, "Ann. de Ch. et de Phys.," 4e sér., t. xi.) We shall, however, omit its discussion.

168. The following table of the specific volumes v of *superheated* steam, as observed by Hirn for centigrade temperatures t, is valuable for technical applications:

p.	$t°$.	v.	p.	$t°$.	v.
1	100°	1.65	3.5	201°	0.6035
"	118.5	1.74	"	225	0.636
"	141	1.85	"	246.5	0.6574
"	148.5	1.87	4	165	0.4822
"	162	1.93	"	200	0.522
"	200	2.08	"	225	0.539
"	205	2.14	"	246.5	0.5752
"	246.5	2.29	5	160	0.3758
2.25	200	0.92	"	200	0.4095
3	200	0.697	"	205	0.414
3.5	196	0.591			

The pressures p are here given in atmospheres.

From the above table of Hirn we construct the following, for the temperature 200°:

p.	v.	pv.
1	2.08	2.08
2.25	0.92	2.07
3	0.697	2.091
4	0.522	2.088
5	0.4095	2.0475

Hence, it appears that steam superheated to 200° gives pv nearly constant, or that it may be supposed to obey the law of Mariotte approximatively.

If, from the data of Hirn, we calculate values for the coefficient of expansion, they are very irregular; thus showing the observations to be affected with probable errors too great for exact theoretical purposes. It is to be hoped, therefore, that before long some other observer, with superior facilities, may repeat these experiments and extend them to the superheated vapours of other liquids beside water.

CHAPTER XI.

ON STEAM-ENGINES, THEIR DEFECTS AND IMPROVEMENT.

169. We shall suppose the reader sufficiently acquainted with the construction and working of condensing and non-condensing engines to dispense with any description of the manner in which the water is turned into steam, which pushes the piston to and fro and does the effective work. Also, that the offices and the technical names of the usual parts of an engine are familiarly known to him. Or if he lack such general or popular information, which in this age every one pretending to be well informed is expected to possess, then we trust that, before attempting to study this chapter, he will acquire it, by reading some descriptive book, and by personal inspection of steam-engines in operation; all of which may be easily accomplished in a few hours in almost any active part of the present civilized world. Consequently, we shall omit descriptions and definitions, and will freely use any technical terms we may have occasion to employ.

170. To any one thus fitly informed, the following brief analysis of the *cycle* of operations constantly recurring in a condensing engine at work will, it is presumed, present no obscurity: 1°, a definite mass or weight of water in the boiler, at the temperature τ and under the corresponding pressure p, is changed into steam of elastic force equal to p, which steam passes into the cylinder and pushes the piston; 2°, the pipe from the boiler being now closed, that steam continues to drive the piston, expanding until it fills the cylinder, in which change the pressure decreases and part of the steam becomes liquefied; 3°, the mixture of steam and water is now driven by the engine out of the cylinder into the refrigerator, where, at the reduced temperature τ_0 and pressure p_0 it becomes entirely liquefied; 4°, lastly, it is, under the pressure p, pushed back by the force-pump into the boiler. And thence again, and in constantly recurring cycles, the same or an equal quantity of water goes through this same series of operations.

IDEAL AND PERFECT ENGINES.

171. Before considering real engines with their imperfections, it is well to form a distinct conception of what would constitute a perfect engine, working between the temperatures τ and τ_0. And for all such engines we have already found Carnot's law,

$$\frac{q - q_0}{q} = \frac{\tau - \tau_0}{\tau}.$$

Let now the physical state of a unit of water in the condenser, or refrigerator, of the temperature τ_0 and under the pressure p_0, be denoted in the diagram of energy by the position a. And suppose a unit of water at a to be driven by the force-pump, with increase of pressure from p_0 to p, and of temperature from τ_0 to τ, into the boiler. Then the curve ab indicates this change, and the

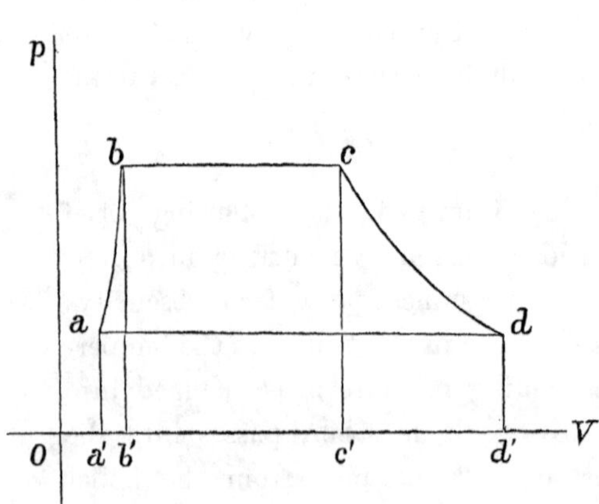

heat absorbed by that unit of water will be

$$\int_{\tau_0}^{\tau} m' d\tau.$$

Next, let the fraction x of that unit of water be turned into steam, under constant temperature τ and pressure p, passing thus by an isothermal change, indicated by the line bc to the expanded condition c and from the boiler into the cylinder. For this change the required heat of vaporization is λx.

The unit mixture of water and saturated steam in the cylinder may now be supposed to pass suddenly, and therefore adiabatically, from the state c to that of d, with decrease of pressure from p to p_0, and of temperature from τ to τ_0; no heat is received or emitted, but in doing this work of adiabatic expansion, part of the steam liquefies, to yield its heat to another part.

Lastly, let negative work of compression and condensation be done by the engine upon this unit mixture, containing now only x_0 of saturated steam, under the constant pressure and temperature p_0 and τ_0; and let its volume be reduced by liquefaction until it is all again liquid, and restored in the condenser, or refrigerator, to a, its original condition. This isothermal change is denoted by the line da of the diagram, and $\lambda_0 x_0$ is the latent heat of condensation due to it.

Adding now these thermal variations, we have

$$\Delta Q = \lambda x - \lambda_0 x_0 + \int_{\tau_0}^{\tau} m' d\tau, \qquad (189)$$

the total variation of heat due to the hypothetical cycle.

172. We have seen (in Art. 98) that for all heat engines the test of perfection is reversibility. Also (in Art. 117) that by the theorem of Carnot,

$$E \frac{\tau - \tau_0}{\tau} dq$$

is the maximum work which can possibly be done by any elementary variation of heat. And therefore, that as, in any perfect engine without waste,

$$E(q - q_0)$$

is the work done, the *dissipation* or waste (equation 125) for a reversible cycle, or perfect engine, is

$$\int \frac{dq}{\tau} = 0.$$

173. If now the ideal engine under consideration be perfect, then its cycle must be reversible, and by differentiation

$$\frac{d(\lambda x)}{\tau} + m' \frac{d\tau}{\tau} = 0; \qquad (190)$$

whence, by definite integration,

$$\frac{\lambda x}{\tau} = \frac{\lambda_0 x_0}{\tau_0} - \int_{\tau_0}^{\tau} m' \frac{d\tau}{\tau}; \qquad (191)$$

and thus for a perfect engine, or reversible cycle, we get as the condition of *no loss* the same equation (177) which, though in a very different way, we have already found for an adiabatic change in saturated steam.

STEAM-ENGINES WORKING IN CYCLES OF CARNOT.

174. Certainly, it is not necessary to prove for steam-engines in particular what has been already shown for all engines supposed to work in cycles of Carnot, that they are perfect. Nor will we probably ever find for such ideal engines a simpler demonstration of their perfection than that of Art. 120; which, by definite integration of the function

$$dq = \tau d\phi,$$

gives the law of Carnot

$$\frac{q - q_0}{q} = \frac{\tau - \tau_0}{\tau},$$

for their efficiency.

175. Nevertheless, as an example of the use of our general formulas for steam, let us apply them to a cycle of Carnot composed of two isothermal cutting two adiabatic lines.

In the physical state represented by the co-ordinates p and v of the point a, let in the boiler a unit of water receive heat sufficient to convert a fraction x thereof into steam; this change will be figured by the isothermal line ab, for which the pressure and temperature are constant. During it, the mixture also passes from the boiler into the working cylinder. And the heat absorbed in this isothermal change is

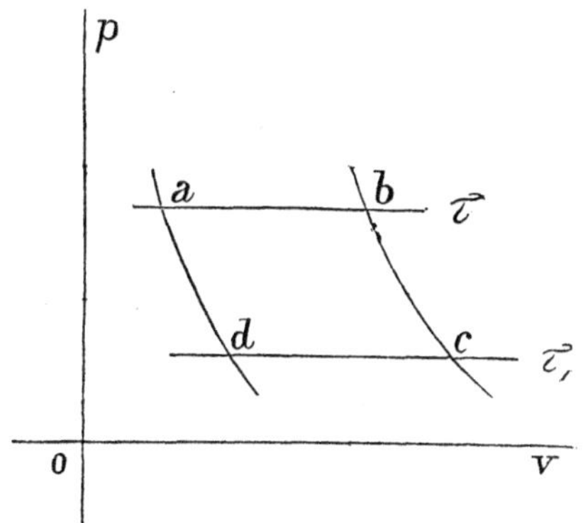

$$q = \lambda x.$$

Next, adiabatically from b to c, there is in the cylinder expansion, but neither loss nor gain of heat; hence the temperature falls from τ to τ_1 and there is partial liquefaction, (Rankine's law); the amount of which, given by the adiabatic equation (177) or (191), will be obtained from

$$\frac{\lambda_1 x_1}{\tau_1} = \frac{\lambda x}{\tau} + \int_{\tau_1}^{\tau} m' \frac{d\tau}{\tau}. \qquad (a)$$

Then from c to d condensation takes place, and the latent heat given out to the refrigerator is

$$\lambda_1 (x_1 - x_0),$$

in which x_1 and x_0 are the values of the variable fraction x at c and at d. And the value of x_0 is determined by the condition, that in the fourth or last change the mixture shall be forced back to its original state a adiabatically.

During this final change, indicated by the adiabatic curve ad, the mixture becomes entirely liquefied; and we have for it, as x becomes nothing at a, the equation

$$\frac{\lambda_1 x_0}{\tau_1} = \int_{\tau_1}^{\tau} m' \frac{d\tau}{\tau}. \qquad (\beta)$$

Subtracting equation (β) from (α) we obtain

$$\frac{\lambda_1}{\tau_1}(x_1 - x_0) = \frac{\lambda}{\tau} x,$$

and, therefore,

$$q = \lambda_1 (x_1 - x_0) = \lambda x \frac{\tau_1}{\tau}.$$

Hence the heat converted into external work in a cycle of Carnot would be equal to

$$q - q_1 = \lambda x \frac{\tau - \tau_1}{\tau},$$

and the efficiency would be

$$\frac{q - q_1}{q} = \frac{\tau - \tau_1}{\tau},$$

as we already know from the the theorem of Carnot for all perfect engines.

An adiabatic change being in a steam-engine a physical impossibility, we pass from the consideration of *ideal* to that of *real* engines, such as are actually employed.

REAL ENGINES.

176. For a first approximation we may suppose the engine without friction or other imperfections, just as is done in mechanics for elementary machines, for the simple pendulum, and for projectiles and falling bodies in vacuo. Also, if the steam expands rapidly in the cylinder, we may assume this change to be adiabatic and the expansion to be complete. The cycle of operations will then be identical with those set forth for an ideal engine in Art. 171; and its diagram, as well as all the formulas, become directly applicable. So that we have for the heat absorbed,

$$q = \lambda x + \int_{\tau_0}^{\tau} m' d\tau ;$$

for that emitted

$$q_0 = \lambda_0 x_0 ;$$

and for that converted into external work

$$q - q_0 = \lambda x - \lambda_0 x_0 + \int_{\tau_0}^{\tau} m' d\tau ,$$

which is identical with (189), as it should be.

We may eliminate x_0 by aid of equation (177) for an adiabatic change, or by the relation

$$\frac{\lambda_0 x_0}{\tau_0} = \frac{\lambda x}{\tau} + \int_{\tau_0}^{\tau} m' \frac{d\tau}{\tau} ,$$

and thus obtain

$$q - q_0 = \lambda x \frac{\tau - \tau_0}{\tau} + \int_{\tau_0}^{\tau} m' \left(\frac{\tau - \tau_0}{\tau} \right) d\tau ; \qquad (192)$$

which is, therefore, the general equation of such engines.

177. Moreover, it is clear that the equations just obtained may be put under the following modified forms, if for approximation the specific heat m', which varies very slightly, be assumed to be equal to c and constant.

The heat received is

$$q = \lambda x + c(\tau - \tau_0); \tag{193}$$

that emitted is

$$q_0 = \lambda_0 x_0.$$

To eliminate x_0, we have

$$\frac{\lambda_0 x_0}{\tau_0} = \frac{\lambda x}{\tau} + c \log \frac{\tau}{\tau_0}.$$

And, therefore, the heat turned into external work is

$$q - q_0 = \lambda x \frac{\tau - \tau_0}{\tau} + c(\tau - \tau_0) - c\tau_0 \log \frac{\tau}{\tau_0}; \tag{194}$$

or

$$\Delta q = \lambda x \frac{\tau - \tau_0}{\tau} + c\left[\tau - \tau_0\left(1 + \log \frac{\tau}{\tau_0}\right)\right]. \tag{194*}$$

178. From the value given for λ by the formula of Regnault (160), and from equations (193) and (194), we may compute the efficiency

$$e = \frac{q - q_0}{q}$$

due to the cycle of such an engine. Making the requisite numerical calculations for the temperatures 150° and 50° centigrade, we find

$$e' = 0.219.$$

Between the same limits a cycle of Carnot gives

$$e = \frac{\tau - \tau_0}{\tau} = 0.236.$$

The difference 0.017 shows the imperfection of the cycle.

179. It is easily proved that the coefficient e', or efficiency of a steam-engine, approximates more and more closely to the maximum e of a perfect engine, or cycle of Carnot, as the *chute de chaleur*, or difference between τ and τ_0, diminishes.

For this purpose, equation (194) gives the following transformations:

$$\tau_0\left(1 + \log\frac{\tau}{\tau_0}\right) = \tau_0 - \tau_0 \log\frac{\tau_0}{\tau},$$

but

$$\tau_0 \log\frac{\tau_0}{\tau} = \tau_0 \log\left(1 - \frac{\tau - \tau_0}{\tau}\right).$$

If we develope the last term by aid of the well-known formula

$$\log(a + x) = \log a + \frac{x}{a} - \frac{x^2}{2a^2} + \text{etc.},$$

making the proper substitutions, and rejecting powers of x higher than the first, we have

$$\tau_0 \log\left(1 - \frac{\tau - \tau_0}{\tau}\right) = -\tau_0\frac{\tau - \tau_0}{\tau}.$$

Hence,

$$c\tau - c\tau_0\left(1 + \log\frac{\tau}{\tau_0}\right) = c\tau - c\tau_0\left(1 + \frac{\tau - \tau_0}{\tau}\right).$$

Or, by reduction,

$$\frac{c}{\tau}(\tau^2 - 2\tau\tau_0 + \tau_0^2) = c\frac{(\tau - \tau_0)^2}{\tau}.$$

Consequently, equation (194) approaches the limit

$$q - q_0 = \lambda x\frac{\tau - \tau_0}{\tau} + c\frac{(\tau - \tau_0)^2}{\tau}.$$

And if we divide this by (193) we get

$$e' = \frac{q - q_0}{q} = \frac{\tau - \tau_0}{\tau};$$

or the limit is a cycle of Carnot.

Geometrically, too, it is evident that as the isothermal line *cd* approaches *ab*, or as τ_0 is nearer to τ, the area of the triangle *ade*, which is the difference of *abcd* and *abce*, or of the cycle of a steam-engine from the cycle of Carnot, does also approach a limit.

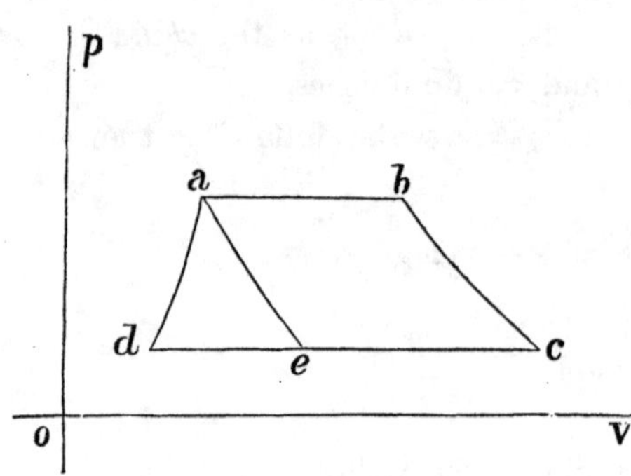

180. If we make in equation (192) the higher temperature equal to τ_1, and change the accent of τ accordingly both in it and in the expression for the heat absorbed,

$$q = \lambda x + \int_{\tau_0}^{\tau_1} m' d\tau.$$

Then dividing (192) by q, the quotient, or efficiency, may readily be reduced algebraically to the form

$$e' = \frac{\tau_1 - \tau_0}{\tau_1} - \frac{\tau_0 \int_{\tau_0}^{\tau_1} m' \left(\frac{1}{\tau} - \frac{1}{\tau_1}\right) d\tau}{\lambda x + \int_{\tau_0}^{\tau_1} m' d\tau}.$$

The second member of which proves that the efficiency is less than that of a cycle of Carnot, expressed by the first term of the second member. Also the value of the last term is a minimum when x is greatest, or unity; which is the case when the steam in the cylinder is not mixed with any liquid portion. There is, therefore, advantage in using dry steam.

181. Such is the importance of the steam-engine, that it is well to consider its theory in various ways. And we shall, therefore, give here another demonstration of its general formula (192); the first was obtained by seeking expressions for the heat received and expended; the following will be based upon changes of energy and work.

The engine being double-acting, its cycle will be that of a single stroke of the piston.

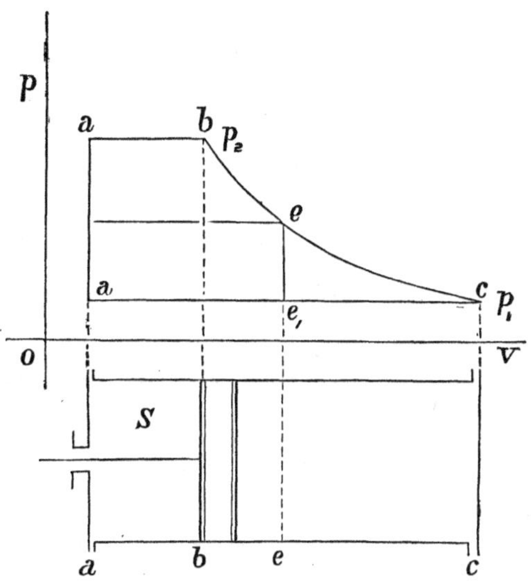

Let x_2 be the proportion of steam in the mixture as it enters the cylinder; and let p_2 and τ_2 be its pressure and temperature. From a to b the steam acts upon the piston with *full head*, or with the pressure p_2; but this pressure is antagonized by the pressure p_1 of the vapour in the condenser, or by that of the atmosphere. If m be the mass of the mixture, and v_2 be its specific volume, then mv_2 is the capacity of the part of the cylinder whose length is ab; and the work due to this part ab of the stroke will be

$$mv_2\,(p_2 - p_1).$$

From b to c expansion takes place, the steam from the boiler being cut off; and at c the specific volume of the mixture will be v_1 and its proportion of steam x_1; this work will be given by the formula (183) already obtained for such changes, or by

$$S = m\left[E\left(\lambda_2 x_2 - \lambda_1 x_1 + \int_1^2 c\,d\tau\right) - p_2 v_2 + p_1 v_1 + \int_1^2 s_0\,dp\right].$$

To eliminate $\lambda_1 x_1$ we have from equation (177) the relation

$$\frac{\lambda_1 x_1}{\tau_1} = \frac{\lambda_2 x_2}{\tau_2} + \int_1^2 c \frac{d\tau}{\tau};$$

making this substitution, and subtracting the back pressure on the piston, which is equal to

$$mp_1 (v_1 - v_2),$$

we get

$$S = m \left\{ E \left[\lambda_2 x_2 \frac{\tau_2 - \tau_1}{\tau_2} + \int_1^2 c \left(\frac{\tau - \tau_1}{\tau} \right) d\tau \right] - v_2 (p_2 - p_1) + \int_1^2 s_0 dp \right\}.$$

And adding to this the work done under full head from a to b, we have for the entire stroke of the piston the work

$$S = m \left\{ E \left[\lambda_2 x_2 \frac{\tau_2 - \tau_1}{\tau_2} + \int_1^2 c \left(\frac{\tau - \tau_1}{\tau} \right) d\tau \right] + \int_1^2 s_0 dp \right\}.$$

From this we must now subtract the work of the force-pump, which at every stroke of the engine takes a mass m from the condenser under the pressure p_1 and forces it into the boiler under the pressure p_2; for which work the value is

$$ms_0 (p_2 - p_1) = \int_1^2 ms_0 dp.$$

Making this reduction, and dividing by the mass m, we obtain for the work due to the unit mixture,

$$S = E \left[\lambda_2 x_2 \frac{\tau_2 - \tau_1}{\tau_2} + \int_{\tau_1}^{\tau_2} c \left(1 - \frac{\tau_1}{\tau} \right) d\tau \right],$$

which is identical with equation (192) already found.

DEFECTIVE EXPANSION.

182. Incomplete expansion is one of the principal defects of steam-engines. In the preceding discussion, we have supposed the expansion from τ_2 to τ_1 to be complete. The table of Clausius (given in Art. 160) shows that between the temperatures 150° and 50° the expansion of saturated steam is 25.7, or nearly twenty-six times its original volume. For such dilation cylinders of enormous size would be requisite. Practically expansion is never pushed to such a degree. The same table shows that between 150° and 100°, it amounts to 3.9 or nearly four, and that between 150° and 75° it exceeds nine. The values of s given in the last column of that table indicate also the relative work for different degrees of expansion.

In the figure of the preceding article (181) let expansion cease at the point e, for which the pressure and temperature are p and τ; then the loss of work or defect due to incomplete expansion is measured by the area of the triangle $ce'e$. At e let also the specific volume of the mixture be v, and let x denote the fraction of steam it contains.

The work of the unit mass during this partial expansion, lessened by that of p_1 the opposing pressure, as given by equation (183) will be

$$E\left(\lambda_2 x_2 - \lambda x + \int_\tau^{\tau_2} cd\tau\right) + \int_p^{p_2} s_0 dp + pv - p_2 v_2 - p_1 (v - v_2).$$

Eliminating λx by its value

$$\lambda x = \lambda_2 x_2 \frac{\tau_2 - \tau}{\tau_2} + \tau \int_\tau^{\tau_2} cd\tau,$$

and integrating, we get

$$E\left[\lambda_2 x_2 \frac{\tau_2 - \tau}{\tau_2} + c\,(\tau_2 - \tau) - c\tau \log \frac{\tau_2}{\tau}\right]$$

$$+ \int_p^{p_2} s_0 dp + pv - p_2 v_2 - p_1 (v - v_2).$$

To this add the work of the full head of steam along *ab*, or

$$v_2 (p_2 - p_1),$$

and from it subtract the energy expended upon the force pump

$$S_0 (p_2 - p_1),$$

and we have after reduction

$$S = (p - p_1)(v - s_0) + E\left[\lambda_2 x_2 \frac{\tau_2 - \tau}{\tau_2} + c\,(\tau_2 - \tau) - c\tau \log \frac{\tau_2}{\tau}\right].$$

From which formula, the data of Regnault and the table of Clausius (Art. 160) enable us to compute values of S and compare the relative efficiency of such an engine with that of either of those already discussed.

If we make the requisite calculations for complete expansion between 150° and 50°, and for partial expansion only between 150° and 100°, and again between 150° and 75°, we find for e' the efficiency, or coefficient, the relative values $0.219 = \frac{2}{9}$, $0164 = \frac{1}{6}$, and $0.205 = \frac{1}{5}$; while a cycle of Carnot gives $0.236 = \frac{1}{4}$ nearly, between 150° and 50° centigrade; and if we compare the values of S for incomplete expansion between 150° and 100°, and for complete expansion from 150° to 50°, we find them to be in the ratio of 99 to 132; thus showing for the former the loss of 33, or one-fourth of the latter.

OTHER IMPERFECTIONS.

183. There are other defects in steam-engines which we will not discuss, for they relate mostly to technical construction and to economical management. We may, therefore, refer for full information concerning them to such works as those of Rankine, De Pambour, Hirn and others; and will here only mention some of them. Of such are wasteful consumption of fuel; incrustation in boilers; obstructed flow of steam; diminution of pressure between the boiler and cylinder, or between the cylinder and condenser; chilling effects of radiation and conduction; friction, etc.

But with all its existing defects, the steam-engine is far from being the very imperfect and faulty machine, which false theoretical ideas have led some to imagine it to be.

The numbers of the last article showing that from 150° to 50° the duty of a *perfect* engine is only 236, and that an engine with its boiler at 150° and condenser at 50°, but with incomplete expansion to 75° only, has a coefficient of 205 or $\frac{1}{5}$ nearly; it follows that 0.93 is the true coefficient for such an imperfect engine. And if of this coefficient four-fifths be available, or effective, then 0.74 would be the actual coefficient of such an engine. To compare this with Hirn's results for his engines, which had a mean coefficient of $\frac{1}{8} = 0.125$, as stated in Art. 10; our engine would have a coefficient of four-fifths of $\frac{1}{5}$ equal to $\frac{1}{6}$ nearly, or 0.205×0.8 equal to 0.164 exactly; and the difference is 0.039, or say four per cent only. Thus we see that allowing 24 per cent for friction and all other defects, Hirn's engines would be perfect.

It also appears that hot air is far from having the comparative advantage over steam, which some lately imagined it to possess; and that, on the contrary, for small ranges of temperature steam is preferable. Indeed, if air like steam could readily be liquefied, and thus used in condensed and less bulky form, such a change would constitute a great improvement for all hot-air engines.

We will now consider some of the theoretical improvements which have been imagined or proposed of late years.

ENGINES OF TWO LIQUIDS.

184. It has been proposed to extend the difference of temperature, or *chute de chaleur*, upon which Carnot's formula

$$e = \frac{\tau - \tau_0}{\tau}$$

shows the maximum efficiency of any heat engine to depend, by combining with a steam-engine, working between the temperature τ and τ_1, another engine using a much more volatile liquid, such as sulphuric ether or benzine, and working between the temperatures τ' and τ_0. The condenser of the steam-engine being thus made to play the part of boiler to the auxiliary ether engine.

It is easily shown that such a double engine is simply equivalent to a single engine working between the extreme temperatures τ and τ_0. For this purpose, each engine may be assumed to be perfect, or to work in a cycle of Carnot. The steam-engine will convert into work

$$Q\frac{\tau - \tau'}{\tau},$$

and it will give to the ether engine

$$Q' = Q\frac{\tau'}{\tau}.$$

This engine utilizes

$$Q' \frac{\tau' - \tau_0}{\tau'},$$

and throws away

$$Q' \frac{\tau_0}{\tau'} = Q \frac{\tau_0}{\tau}.$$

It is clear that the heat used by the double engine will be

$$Q - Q \frac{\tau_0}{\tau} = Q \frac{\tau - \tau_0}{\tau},$$

or its coefficient is the same as that of a single engine working between the extreme temperatures τ and τ_0; which was to be proved.

An engine with two fluids may, therefore, be used to extend the *chute de chaleur*.

Not only ether, but other volatile liquids, *e. g.*, chloroform and chloride of carbon, have been tried. Such engines have been used in France, and one invented by Du Trembley was in 1855 tried in the steamer "Brésil" with considerable economy of fuel.

They have been abandoned chiefly, perhaps, for fear of fire; though escape of noxious fumes, corrosion of metal, etc., are other objections urged against them.

STEAM-JACKETS.

185. Around the cylinder of his engines Watt placed another cylinder of larger diameter, with a space between them filled with steam from the boiler; and this contrivance is called a steam-jacket.

It is not known what led him to adopt such an arrangement. It has since been criticised and condemned hastily as a very faulty way of preventing radiation and conduction; which it was alleged could be better done by an envelope of wood, or other non-conducting material. Such ignorant stupidity cannot be ascribed to that illustrious man.

In the locomotive engine of Mr. Stephenson, the cylinder was placed in the flue from the furnace, for economy of heat and fuel.

And it has been practically found by Hirn that an economy of not less than 20 per cent may be realized from jacketed engines.

Very different views have been entertained in theoretical explanation of this important economy or advantage, which practical results prove to be unquestionable as a fact. And clearly it has nothing to do with radiation and conduction, which take place from the outer surface of the jacket even to a greater degree than could happen for the working cylinder of smaller size.

The important discovery of Rankine, that liquefaction takes place in a cylinder working expansively, has induced some to adopt the hypothesis that a steam-jacket imparts additional heat, prevents liquefaction, keeps the steam saturated but not superheated, and thus causes the increased economy attested by experience.

Our equations will verify the truth of that hypothesis. The heat required first to heat a unit of water and then to convert it into steam, is that called by Regnault its *total heat;* and it must be increased by the quantity from the jacket preventing liquefaction. The whole quantity received is, therefore,

$$Q = m'(\tau - \tau_0) + \lambda - \int_{\tau_0}^{\tau} m d\tau.$$

And the heat lost is

$$Q_0 = \lambda_0.$$

Hence, that converted into work is

$$\lambda - \lambda_0 + m'(\tau - \tau_0) - \int_{\tau_0}^{\tau} m d\tau.$$

In these expressions the last term is negative, because (Art. 149) the specific heat m of saturated steam is negative. But the equation of Sir W. Thomson (167) gives

$$m' - m = \frac{\lambda}{\tau} - \frac{d\lambda}{d\tau};$$

and therefore, by substitution,

$$Q - Q_0 = \int_{\tau_0}^{\tau} \lambda \frac{d\tau}{\tau}.$$

By the formula of Regnault,

$$\lambda = 606.5 - 0.695t = 796.25 - 0.695\tau,$$

and

$$\frac{\lambda}{\tau} = \frac{796.25}{\tau} - 0.695.$$

whence

$$Q - Q_0 = 796.25 \log \frac{\tau}{\tau_0} - 0.695 (\tau - \tau_0).$$

And if we apply this formula to an engine working between 150° and 50°, we find

$$Q - Q_0 = 144.4.$$

But for an ordinary engine with complete expansion between the same temperatures we found (Art. 182) the work equal to 132 and the efficiency 0.219; there is, therefore, a gain in work of one-eleventh, or nine per cent nearly. But the amount of heat received was greater to the extent of

$$\int_{\tau_0}^{\tau} m d\tau,$$

or 114.45 calories; and the economical coefficient is, therefore, 0.201 only. The cycle is, consequently, more imperfect than that of an ordinary engine with incomplete expansion to 75°, for which 0.205 is the efficiency.

What is the interpretation of these apparently discordant results, an actual gain of nine per cent of work and yet a smaller coefficient? Evidently, the additional heat comes only from the fire; and hence there may be more work but less economy, if that additional heat be not such as would be otherwise lost or wasted. If the cylinder be put in the flue or chimney of the furnace, as in the locomotive of Mr. Stephenson, or the heated gases of combustion pass into a hot-air jacket before escaping, then evidently any additional heat absorbed would be a positive economy, as well as a gain, which would otherwise be lost.

But if, as in the steam-jacket of Watt, experience shows both gain and economy amounting, according to Hirn, even to 20 per cent; then such an advantage is not at all explained or accounted for by the hypothesis that the latent heat of the steam in the jacket keeps that in the working cylinder saturated but without liquefaction.

Accordingly, we find that while Rankine adopts that hypothesis, it is disputed and rejected by others. Combes ascribes all advantage to the influence of the walls of the cylinder, which, chilled by expansion and condensation, determine, at each introduction of steam, the immediate liquefaction of a part which comes into contact with them. And Rankine mentions that in double-cylinder engines, where the expansion begins in a small and finishes in a large cylinder, if the small cylinder only be jacketed, the liquefaction is prevented almost entirely.

Moreover, if the piston move rapidly, expansion would be nearly adiabatic, sufficient time not being given for steam in the jacket to supply heat to the non-conducting steam in the cylinder. This whole subject seems, therefore, to demand further experimental investigation for its elucidation.

SUPERHEATED STEAM.

186. With increase of temperature, the expansive force of steam varies so rapidly, that it is said to vary in geometric progression when the former varies only arithmetically; and the empirical formulas, which have been proposed for it, are logarithmic or exponential. For a pressure of ten atmospheres, used in locomotives and other engines, the corresponding temperature is about 180°; and that pressure cannot be much exceeded without great danger of explosion. The *chute de chaleur*, or difference $(\tau - \tau_0)$, upon which the duty of an engine depends, cannot, therefore, be much increased by making the boiler hotter.

It is, however, quite practicable to elevate the temperature by letting the steam from the boiler pass to the cylinder through a pipe, or system of pipes, exposed to the heating action of contact with the hot escaping gases of combustion in the flue of the furnace. And it is clear that in

$$\frac{Q - Q_0}{Q} = \frac{\tau - \tau_0}{\tau} = 1 - \frac{\tau_0}{\tau}$$

the fractional part of the lost heat, measured by the last term, varies directly as τ_0 the lower, and inversely as τ the higher limit.

As the hot gases in the flue escape into the atmosphere, whatever heat can be taken from them to *superheat* the steam is obviously only so much saved or economized. And without increased consumption of fuel a decided gain is thus realized.

To calculate that gain, we must add to the total heat absorbed by a unit mass in an ordinary engine, and expressed as above by

$$\lambda_1 + \int_{\tau_0}^{\tau_1} c d\tau,$$

or by the formula (160) of Regnault, the additional quantity needed to heat it still higher to τ, which gives for the heat received

$$Q = \lambda_1 + \int_{\tau_0}^{\tau_1} c d\tau + \int_{\tau_1}^{\tau} m d\tau,$$

in which m is the specific heat of superheated steam, or of steam-gas, as Rankine calls it.

The lost heat given to the condenser is $\lambda_0 x_0$; and that used in work is

$$Q - Q_0 = \lambda_1 - \lambda_0 x_0 + \int_{\tau_0}^{\tau_1} c d\tau + \int_{\tau_1}^{\tau} m d\tau.$$

From this we may eliminate x_0 by the following considerations. The general thermodynamic function

$$\phi = \int \frac{dQ}{\tau}$$

expresses any determinate state or physical condition of a given substance. If, therefore, in a steam-engine, by any cycle of operations, a mass of water pass back into its initial state of pressure, density and temperature, then its first and last states are identical, and for the entire cycle

$$\int \frac{dQ}{\tau} = \phi_0 - \phi_0 = 0.$$

Applying this to our engine, we have

$$\frac{\lambda_1}{\tau_1} + \int_{\tau_0}^{\tau_1} c \frac{d\tau}{\tau} + \int_{\tau_1}^{\tau} m \frac{d\tau}{\tau} - \frac{\lambda_0 x_0}{\tau_0} = 0.$$

And substituting the value of $\lambda_0 x_0$ given by this equation in the first, we have

$$Q - Q_0 = AS = \lambda_1 \frac{\tau_1 - \tau_0}{\tau_1} + \int_{\tau_0}^{\tau_1} c \left(1 - \frac{\tau_0}{\tau}\right) d\tau + \int_{\tau_1}^{\tau} m \left(1 - \frac{\tau_0}{\tau}\right) d\tau.$$

But for an ordinary engine we have found (Art. 181) that

$$Q - Q_0 = AS_0 = \lambda_1 \frac{\tau_1 - \tau_0}{\tau_1} + \int_{\tau_0}^{\tau_1} c \left(1 - \frac{\tau_0}{\tau}\right) d.$$

Consequently,

$$A\,(S - S_0) = \int_{\tau_1}^{\tau} m \left(1 - \frac{\tau_0}{\tau}\right) d\tau \qquad (195)$$

is the economy or gain in work.

The numerical calculations and the integration demanded by this formula are easily performed, if for m we employ its mean and constant value, found by Regnault equal to 0.48 nearly. An engine whose boiler is at 150°, and whose condenser is at 50°, gives for steam superheated to 300°,

$$x_0 = 0.771, \qquad \text{and} \qquad AS = 156\,;$$

and an ordinary engine gives (Art. 182) for like temperatures 132 thermal units. The gain would, therefore, be 24 units, or nearly eighteen per cent, with the same consumption of fuel.

Various arrangements for superheating steam have been tried; one of the more ingenious of which is that of Mr. Wethered, which he call "*combined steam;*" but we must refer to technical treatises for descriptions of the construction of engines and other details.

NOMINAL AND ACTUAL HORSE-POWER OF STEAM-ENGINES.

187. When steam was first used instead of horses to pump water from mines, it became necessary to compare the power of an engine with that of horses. Various estimates of what a strong horse can do were made by different engineers. But finally the work done in elevating 33,000 pounds through a foot in a minute of time was generally adopted in Great Britain and in this country, as the conventional measure of one horse-power in steam-engines.

This is also equivalent to the work of raising 550 pounds one foot per second; which mode of calculation is sometimes more convenient.

In France a slightly different usage prevails, the horse-power, or *cheval-vapeur*, being there assumed to be equal only to 75 kilogrammètres, or French units of work, per second. To agree with the British unit, it should be 76 kilogrammètres; but except in comparative theoretical calculations the difference is so small that it may be technically disregarded.

We should observe the necessity existing to take time into account in measuring the efficiency of machines and animals. The same given quantity of work can be done by a child, a man, or a horse, but the horse will do it more rapidly; hence time must be reckoned in the comparison. The discussion of machines, given in Art. 22, and their law

$$\Sigma \int_1^2 P\delta p = \Sigma \frac{m}{2}(v_2^2 - v_1^2),$$

show that velocity enters into such calculations. And in the same article it is also shown that when the power and resistance become equal, and the velocity constant, the machine works with maximum economy and advantage.

Supposing now, for a steam-engine, that constant velocity and equality of action and reaction established, or that its cycle of operations has become periodically constant, we readily see that the rule used for computing the actual horse-power of an engine is correct. Which rule is as follows: Multiply the total pressure on the piston by the length of its stroke in feet and this by the number of strokes per minute in both directions, then divide that product by 33000 for British measures. For French measures the kilogrammètre per second is the unit of work, the metre is the linear unit, and 75 is the divisor.

Algebraically the work per minute will be

$$S = \pi r^2 lnp = \int_1^2 pdv,$$

in which l denotes the length of the stroke, n the number of strokes per minute, p the average effective pressure upon the unit of surface, πr^2 the area of the piston, and v the volume developed.

The horse-power given by the above rule and definition will therefore be for British usage,

$$H = \frac{S}{33,000}; \tag{196}$$

and for French measures and units of work

$$H' = \frac{S'}{60 \times 75} = \frac{S'}{4500}. \tag{197}$$

To obtain in French measures the work S', if p be the mean pressure in atmospheres, and v the volume developed by one stroke of the piston, then

$$S' = 10330npv, \tag{198}$$

reckoned in kilogrammètres. The coefficient 10330 being the atmospheric pressure upon a square metre.

188. The formulas just given would be perfectly correct and of easy application, if it were possible to determine exactly the mean effective pressure p upon the piston. It is evidently only the resultant, or difference, of the direct pressure of the steam upon one face of the piston, and of the counter pressure of the vapour in the condenser, or of the atmosphere for non-condensing engines, upon the other face. Moreover it is always exactly equal to and varies with the intensity of the resistance or work; which itself is composed of two parts, that of the load or useful work, and that due to friction and other opposing actions in the engine itself.

In coming from the boiler into the cylinder the steam is obstructed by the pipes and valves, and its pressure is thereby somewhat reduced. This obstruction the manufacturer may and should try to diminish by using short and large pipes, with open valves; but the difference of pressure between the boiler and the cylinder is chiefly due to the fact that in the cylinder the piston, being moveable, acts as if it were a safety valve, and lowers the pressure to the amount requisite to equalize its intensity with that of the resistance, as was first proved by De Pambour.

The back pressure also cannot be determined, for it varies largely, owing to more or less obstruction to the escape of the steam from the cylinder into the air or condenser, to the mixture of air and water with that steam, and to other causes not yet sufficiently investigated.

By trials made upon various engines, Gen. Morin has sought to obtain, for the difference of pressure in the boiler and cylinder, a coefficient of reduction, which for locomotives he makes equal to 0.60; for steamers 0.80; and for stationary engines 0.85 to 0.90. But De Pambour has proved, by careful trials, that even for the same engine these coefficients are not constant, but variable with its work or load.

Under these circumstances, the actual mean value of p can be practically best determined by using the diagrams drawn by the indicator of Watt, and recording the corresponding number of strokes of the piston by an automatic register.

De Pambour gives, however, a formula for determining the value of p from the quantity of water evaporated in the boiler, or supplied by the feed pump, which he deduces as follows: let m be the quantity of water evaporated in the unit of time and s the specific volume of the steam it produces in the boiler under the pressure p'; it passes into the cylinder where its pressure will reduce to the effective pressure p equal to the resistance R, and by Mariotte's law its volume will be

$$ms\frac{p'}{R}.$$

Again the area of the piston being a and its length of stroke l, their product is the volume of the cylinder v, and nv will be the quantity of steam used by n strokes in the unit of time ; hence the equality of demand and supply gives

$$nv = ms\frac{p'}{R};$$

whence

$$R = \frac{ms}{nv}p'.$$

Unfortunately, however, for De Pambour's attempt, it was not then known that steam becomes liquefied when expansion occurs in the cylinder, and Mariotte's law fails entirely.

Neither the method of *coefficients* of Morin, nor that of De Pambour, can be used for such calculations ; and both must be abandoned.

189. Hence it appears that as the actual work of an engine varies with its load, its *horse-power*, which only measures that work, is in fact variable and indeterminate. This truth seems to have been well known to Watt and to have induced him, as a constructor, to adopt for his engines a purely fictitious, or *nominal*, horse-power still used in trade. And which is calculated thus: Multiply the fictitious pressure 7, by the assumed velocity 128 feet per minute, by the area of the piston in inches and by the cube root of the stroke in feet, then divide by 33.000. Or algebraically

$$H = \pi r^2 \frac{7 \times 128}{33,000} \sqrt[3]{l}.$$

Which reduces, if we make π equal to twenty-two sevenths, to

$$H = r^2 \frac{64}{750} \sqrt[3]{l}.$$

The British Admiralty adopt a different rule: Multiply the fictitious pressure 7 by the actual velocity of the piston in feet and by the area of the piston in inches, and divide the product by 33,000 ; this gives

$$H = \pi \frac{7}{33000} nlr^2 = \frac{nlr^2}{1500}.$$

In the French marine the rule for the *nominal* horse-power is simpler. If p be the assumed effective pressure, and v the velocity of the piston per second, then

$$H = \frac{pv}{75}.$$

And from this formula another much used by French constructors is derived. In an engine of Watt assumed to work with a pressure of one atmosphere, and with the velocity of one metre per second, v becomes unity, and therefore

$$H = \frac{p}{75}.$$

If now we substitute for p its value, equation (198), then

$$H = \pi \frac{10000}{4 \times 75} (2r)^2 = 100\,(2r)^2.$$

In which $(2r)$ the diameter of the piston is measured in metres; but if we reduce it into decimètres

$$H = (2r)^2. \qquad (199)$$

Or we have the extremely simple rule that: *the square of the diameter of the piston in decimètres is the nominal horse-power.*

And from this, for any other similar engine working with the pressure p and velocity v, the nominal horse-power is

$$H = pv\,(2r)^2. \tag{200}$$

Or we have the rule: *Multiply the pressure by the velocity and by the square of the diameter of the piston in decimètres.*

To these *nominal* horse-powers, it is customary to apply coefficients of reduction, varying from 80 to 85 per cent, for defects of construction; and to multiply these again by the factors of reduction 0.60 to 0.90 for difference of pressure between the cylinder and the boiler; thus giving as limits 0.50 to 0.75, called by Morin *factors of construction.*

190. The idea has been entertained by constructors, habituated to the old formulas, that those required by the new theory of heat are much more complex; this is far from being true; and it is much to be regretted that any such false notion should be allowed to perpetuate the use of erroneous methods, which can scarcely fail to retard progress and improvement.

The maximum effect which a given quantity of heat Q can produce in a *perfect* engine is very easily computed from the formula of a cycle of Carnot,

$$S = EQ\frac{\tau - \tau_0}{\tau}.$$

And having thus obtained the *duty* of a *perfect* engine, for any given limits of temperature, we may with great facility compare with it the work done by any other engine; using for that purpose the formulas we have demonstrated; which also may often be much simplified by using tables and approximations sufficiently exact for technical purposes.

Moreover, the law of action and reaction gives for the relation between the power applied and the resistances overcome that their sum will be equal to zero, or

$$S = R + r + r' + r'' + r''' + \text{etc.}$$

It is, therefore, only necessary to determine successively the separate hurtful resistances and to subtract them from S, in order to obtain R, the useful work. And it is difficult to conceive of any theory which could lead to calculations more simple.

191. We now quit this most important and interesting of the applications of our theory, regretting much that the necessary limits of an elementary treatise do not let us give more information of practical details; but thc reader will find them in the numerous valuable books on the construction and management of engines, which are written by experienced and able professional men, from time to time, as perpetual progress demands.

CHAPTER XII.

MISCELLANEOUS.

192. To any one who has followed step by step the rigid chain of inductive reasoning presented in the preceding chapters, from which every supposition has been most carefully excluded, except the single hypothesis that heat and other forms of energy are convertible and indestructible, it will now be evident that, in all our knowledge of the laws of the physical world, there are none established upon a firmer basis than the two general propositions of Joule and Carnot; which constitute the fundamental laws of our subject, and which may both be combined in the single expression

$$\phi = \int \frac{dQ}{\tau}.$$

But their applications are innumerable; and when we reflect how generally physical phenomena are connected with thermal changes and relations, it at once becomes obvious that there are few, if any, branches of natural science which are not more or less dependent upon the great truths under consideration. Nor should it, therefore, be matter for surprise that already, in the short space of time, not yet a generation, elapsed since the mechanical theory of heat has been freely adopted, whole branches of physical science have been revolutionized by it.

All we propose to do in the remainder of this work, all that can be done in the compass of one volume, is to present a selected few of the more interesting general applications; in number and in variety, however, sufficient to give a somewhat adequate idea of their extent and utility.

PHYSIOLOGICAL.

193. The body of an animal, not less than a steamer, or a locomotive, is truly a heat engine, and the consumption of food in the one is precisely analogous to the burning of fuel in the other. In both the chemical process is the same, that called combustion.

To the illustrious but sadly unfortunate Lavoisier, the great founder of modern chemical science, we are indebted for the discovery that combustion is generally oxidation. The word *generally* is here used only because sulphur, chlorine, etc., play the part of substitutes for oxygen in combustion.

To Lavoisier, too, we owe the explanation of respiration, in which inhaled oxygen is perpetually exchanged for ejected carbonic acid and vapour of water, products of combustion.

Other organs aid the lungs in the constant cycle of nutrition and excretion; the skin, for instance, emitting perspiration, and the kidneys urine; while the mouth, the stomach, and the intestines replenish waste.

These facts are so familiar to all that no one need demand further proof than his personal knowledge of the general truth that the human body is a heat engine consuming food, water, and air.

But when we seek to trace that truth in all its minor details, difficulties arise, like those which present themselves to a person ignorant of the construction of a locomotive, conscious by crude observation that its activity is due to fire and steam, yet quite unable to see just what are the offices of many of its organs or parts.

So that even to the best informed physiologist obscurities exist which cannot be comprehended.

Hence objections to Lavoisier's theory have been from time to time urged. Among such objections we may allude to that which finds the heat of the body generally and uniformly diffused instead of being concentrated in the chest. But this is easily answered by regarding arterial blood as oxygenated, and the combustion as therefore taking place in every part of the body; from which to the lungs the veins carry back blood in a carbonated state.

A more serious objection has been adduced, that friction, especially in the smaller blood-vessels, must develope heat. Without doubt, animal heat is thus in part produced. But whence the expenditure of energy causing that friction, and which must be itself accounted for?

Precisely here the mechanical theory of heat comes to our relief. The power which drives the blood through its vessels is the muscular exertion of the heart,—a force-pump to receive blood from and send it to every part of the body, the discovery of Harvey,—darkly, yet certainly, we therefore see that the rhythmic pulsations of the heart, like those of the piston of an engine, are dependent upon and consequently due to that cycle of nutrition and excretion which sustains physical or organic life. But precise knowledge of the involuntary action of the heart will probably ever be denied us; for even when a muscle acts voluntarily, we can form no conception of how mind thus subjects matter to volition; and the union of mind, or nervous agency, with matter is an impenetrable mystery. Yet, even in this obscurity, we know that all which is not spiritual and immortal in our being is either matter or energy; neither of which can be created or destroyed, except by their Divine Author; nor continually expended without exhaustion or replenishment.

Directly then, or indirectly, the chemical theory of Lavoisier accounts not only for animal heat, but also for all the complex phenomena united in what is called organic or physical life; and our bodies do literally burn out of their earthly existence, both before and after death, decay being only slow combustion.

That matter and force constitute all which is physical, and that they cannot be created or destroyed, except by God who made them, is in few words the enunciation of the law of conservation of energy. And everything physical being subject thereto, it follows that no physiological action can take place except with expenditure of energy derived from food; also, that an animal performing mechanical work must from the same quantity of food generate less heat than one abstaining from exertion, the difference being precisely the heat-equivalent of that work.

These views, which both amend and complete those of Lavoisier, were first briefly published by Joule in 1843, but soon afterwards they were much more fully and perfectly set forth in Germany by Mayer, who aptly likened the agency of mind, or will, in voluntary motion, to that of the helmsman who steers but adds nothing to the force which drives a steamer.

They were first verified experimentally in 1858 by Hirn, who enclosed men in a tight wooden box, large enough to let them work on a treadmill, elevating their own bodies as the work done; air in measured quantities was introduced and conducted off in tubes; and both the heat emitted and the carbonic acid exhaled during a given time were carefully determined, alike when the man was at rest and when he was at work.

The ratio of the heat emitted to the carbonic acid generated was found much greater for repose than for periods of work; thus proving conversion of heat into work.

It should here be remarked that these experiments do not at all conflict, but on the contrary do perfectly accord, with the result of common experience, that muscular exercise increases respiration and temperature; the numerical data of Hirn so show, but they also *demonstrate* that the *ratio* of the heat emitted to the carbon consumed is less when part thereof is expended in work and thus ceases to exist as heat.

Hirn sought to calculate the mechanical equivalent of heat from such experiments; but for that exact purpose they lack the requisite precision. From them, however, Helmholtz has since ingeniously deduced the economical coefficient of the human body regarded as a heat engine. He observed that the quantity of heat given off in an hour by a man in repose, as proved by Hirn, is just what would do the work of elevating his body to a height of 540 metres. Also that this 540 metres is the height to which a man climbs up a mountain in an hour. But Hirn found the amount of respiration five times as great when a man is climbing. Hence, it follows that *one-fifth* is the available or economical coefficient.

From the force of the blood in the arteries, Helmholtz also calculates that the heart would elevate its own weight in an hour through the height of 6670 metres. And as in mountainous regions the strongest locomotive can only ascend about 800 metres per hour, Helmholtz finds the heart as a machine eight times as effective.

Others have since attempted to perfect the interesting investigations thus begun by Hirn. Among them Beclard tried to determine the heat developed in the muscles of the arm by means of common thermometers ; and it is easily shown not only that he reasoned imperfectly, but also that the variation of temperature must be far too small to be indicated by a common thermometer. His experiments, therefore, were of use chiefly in drawing attention to the necessity of studying the action, not in the body as a whole as done by Hirn, but in particular muscles, and with the aid of much more refined instrumental means of measurement.

Accordingly, the delicate thermoelectric multiplier of Melloni has since been employed in some experiments made upon frogs. But the results are only interesting in that they show the phenomenon to be much more complex than was at first imagined ; that lactic acid is produced in muscular action ; and that this chemical change in the muscle itself interferes with the purely thermal effects.

LIQUEFACTION AND REGELATION.

194. All the general formulas deduced from our fundamental laws, without hypothesis except that heat is energy, are applicable not merely to vaporization, or steam, but also to liquefaction and every other thermodynamic change which may be regarded as a function of the pressure, density and temperature of any substance. This has been already stated in Art. 147, and to some peculiar phenomena attention has been drawn. But it may be desirable, and certainly will be instructive, to discuss more fully a few such facts.

The formula of Sir W. Thomson, or theorem of Carnot,

$$l = A\tau \frac{dp}{dt},$$

gives for all bodies that of Clapeyron,

$$\int_{v_0}^{v} l dv = A\tau (v - v_0) \frac{dp}{dt}.$$

Whenever a body melts, it absorbs heat from without; and the first member of this equation is, therefore, always positive. Consequently, the second member must be so too. Hence the two factors, in the product

$$(v - v_0) \frac{dp}{dt},$$

are always either both positive, or both negative. Whenever, therefore, melting causes increase of size, the pressure and temperature will increase or diminish together. But if, as in case of ice, melting lessens the volume, then increased pressure will be attended with decrease of temperature, or the freezing point will be lowered by compression.

This conclusion, though first theoretically deduced by his brother, was experimentally verified, as stated in Art. 147, by

Sir W. Thomson; who, for this purpose, subjected a mixture of ice and water, with a delicate thermometer, to compression in the apparatus of Oersted for measuring the compressibility of liquids, which is described in almost every treatise on experimental physics. Thus he obtained, for pressures of 8.1 and 16.8 atmospheres, the depressions 0°.059 and 0°.129 in the temperature of the melting point. From which we have the proportionate reduction 0°.0075 for one atmosphere.

To compare this with theory, denoting by λ the latent heat of liquefaction, equal to 79.25 thermal units, we have

$$\frac{d\tau}{dp} = A\frac{\tau}{\lambda}(v - v_0).$$

Also $\tau = 273$; the volume v of the unit, or kilogramme, of water is a litre, or 0.001; and for ice v_0 will be 0.00108. Hence

$$\frac{d\tau}{dp} = \frac{273\,(0.001 - 0.00108)}{425 \times 79.25}.$$

As an atmosphere is equal to the pressure of 10334 kilogrammes per square metre,

$$p = 10334p',$$

and

$$\frac{d\tau}{dp'} = 10334\,\frac{273\,(-\,0.00008)}{425 \times 79.25} = -\,0.0070.$$

Hence the accordance between the observed and calculated results is as close as could be desired.

195. With the apparatus of Oersted, made as usually of glass, it was not possible for Sir W. Thomson to try very powerful pressures; but Mousson has since done so by using a similar but very strong instrument of iron. A detached piece of iron is put in it, and the rest of the cavity is filled with water, which is afterwards frozen. By means of the compressing screw, an enormous pressure is then brought into action and the ice is thereby liquefied; which is known by the piece of iron within becoming free to fall from one part of the vessel to another. In this way Mousson lowered the melting point of ice even to — 15° centigrade, or to 5° Fahrenheit; but the corresponding pressures were not determined.

Analogous experiments have been tried by Bunsen upon paraffine and spermaceti, compressed by a mercurial column in the short leg of a bent glass tube, like that of Mariotte; and the results show elevation of the melting point with increase of pressure, as they should do for these substances.

196. The most interesting facts in relation to this subject are those to which Faraday has applied the word *regelation;* and to which in 1850 he drew attention as "a remarkable property of ice in solidifying water in contact with it. Two pieces of moist ice will consolidate into one. Hence, the property of damp snow to become compacted into a snow-ball, an effect which cannot be produced on dry, hard-frozen snow. A film of water freezing when placed between two sets of icy particles, though not affected by a single set. Certain solid substances, as flannel, will also freeze to an icy surface, though others, as gold-leaf, cannot be made to do so." This fact, thus described by Faraday, is easily verified by squeezing lumps of ice together with the fingers.

At the meeting of the British Association in 1857, the true explanation of *regelation*, about which others had speculated, was given by Prof. J. Thomson, and we will quote his own words: "pieces of ice, on being pressed together at their points of contact, will at those places, by virtue of the pressure, be in part liquefied and reduced in temperature; and the cold evolved in their liquefaction will cause some of the liquid film intervening between them to freeze. It is thus evident that by continued pressure fragmentary masses of ice may be moulded into a continuous mass; and a sufficient reason is afforded for the reunion, found to occur in glaciers, of the fragments resulting from an ice cascade, and for the mending of the crevasses or deep fissures which result occasionally from their motion along their uneven beds.

"The liquefied portions being subject to squeezing of the compressed parts in which they originate, will spread themselves out through the pores of the general mass, by dispersion from the regions of greatest to those of least fluid pressure. Thus the pressure is relieved from those portions in which the compression and liquefaction of the ice had set in, accompanied by the lowering of temperature. On the removal of the cause of liquidity, namely the pressure, the cold evolved in the compressed parts of the ice and water freezes the water again in new positions, and thus a change of form or plastic yielding of the mass of ice to the applied pressure occurs.

"Ice is thus shown to be incapable of opposing permanent resistance to pressures, and to be subject to gradual changes of form while they act on it; or in other words, it is shown to be possessed of the quality of plasticity."

A very simple and beautiful experiment has been contrived by Tyndall for the verification and illustration of the above views of Thomson.

In a hollow spherical mould, made of dense wood, a lump of ice is subjected to powerful compression; which first crushes it into small fragments, and then by continued pressure first partially liquefies and then freezes the ice again into a solid, clear, and transparent ball.

Also, Tyndall, in his investigations of the phenomena of glaciers, so well studied by him among the Alps, has applied the theory of *regelation* to their explanation, and has described them in a manner so admirable as to make the whole subject not only his own, but one of the most interesting of modern physical researches.

STABILITY OF WATER.

197. Many phenomena prove that the particles of liquids cohere powerfully. And as the experiments of Donny, Dufour, and others show that water may be heated above its ordinary boiling point without becoming steam, or cooled below its usual freezing point without forming ice, neither of which changes could take place without increase of size, it follows that a very strong cohesion tends to keep it liquid. Moreover, its very slight compressibility proves its fluidity to be only the property of tangential displacement, its particles merely gliding upon each other.

In Art. 142 it is shown that the latent heat of vaporization, or expenditure of energy, requisite to change a liquid into vapour, is a maximum when it takes place normally. And precisely the same mode of reasoning and formulas may be used to prove that the heat evolved in freezing is a maximum when it occurs normally.

Other bodies possess like properties; thus sulphur, phosphorus, etc., may be cooled below their melting points without solidifying; and sulphuric acid, camphine, caoutchoucine, etc., boil explosively. Analogous phenomena present themselves also in the anomalous retardation of the crystallization of supersaturated solutions of glauber salt. But for these the thermal changes have not yet been sufficiently investigated.

DISSOCIATION.

198. This word *dissociation,* now technically used to signify chemical decomposition by heat, is here given only because it is so used. A better one, perhaps, is *thermolysis,* analogous to *electrolysis.*

Yet, as geographers give names to countries they are the first to visit, astronomers to new planets, and chemists to new metals and compounds, so Deville, the pioneer in this field of discovery, one of the richest to which chemists have given attention since the day of Lavoisier, has named it *dissociation;* and the word is likely, therefore, to be a permanent fixture in science.

It is also well to give new names to new subjects of thought and investigation, lest they be contemned and ignored, if not deemed worthy even of a name. In itself a word is of little consequence, but its meaning should be definite.

The fact that platinum, when heated to its melting point, decomposes water into hydrogen and oxygen, without chemical union with either, was first experimentally observed by Grove. That experiment was repeated by Deville, who founded upon it and other analogous facts his grand discovery that heat alone can decompose all chemical compounds, and in doing so acts always in a perfectly definite manner. Its action being in fact in strict accordance with thermodynamic laws, and capable of being expressed by the very same general formulas which we have given for steam and other physical phenomena.

Mathematically, the discovery of Deville may be enunciated as follows: the physical state of any substance may be always expressed by a function

$$\phi\ (pvt) = 0,$$

which, though unknown, is yet always determinate. And, therefore, the pressure, density, and temperature are variables whose particular values control all such changes as the liquefaction of solids, the vaporization of liquids, and the chemical decomposition of such vapours, if compound, into their constituent elements. Moreover, one single law or function embraces all these changes.

The importance of such a discovery cannot fail to be appreciated by any one even slightly acquainted with chemical science. For valuable, beautiful, and attractive as are its facts, unfortunately they are simply innumerable. Human life is not long enough to learn them. No memory could retain them. No general laws embrace and explain them. Fascinated with their beauty and utility, baffled by their multiplicity, we weary of the patient, endless toil they demand almost in vain. Even the atomic hypothesis of Dalton loses its simplicity for organic substances, and attempts at general theory, or even classification, end in hopeless confusion. In this I am unconscious of exaggeration; certainly no disparagement is intended. If the sketch be even approximatively true, how valuable any discovery which gives promise of bringing particular facts under general law, or of introducing those mathematical methods which are of such service in other branches of physical science. But we must quit these reflections, and present the reader the facts of this subject.

199. Dissociation was first studied by H. St. C. Deville for water, hydrochloric, sulphurous and carbonic acids, and for carbonic oxide. Others have since extended his researches, and all compounds are now believed to obey the same laws.

We select, as an example, clear, definite, and practically important, the action of heat upon pure carbonate of lime, studied by Debray.

In the year 1750, the chemical difference of lime and limestone was discovered by Dr. Black, who extended the inquiry to the caustic and carbonated alkalies, and called carbonic acid gas *fixed air*, because found in solid combination. But from that day until recently, when Debray resumed the investigation, all were content to regard the phenomena which occur when carbonate of lime is heated in kilns, or open vessels, as alone needing attention. Thus, for a century, no one seems to have thought it worth while to heat that substance in close vessels, or under pressure, with the view of observing the difference, if any, in the results.

In few words, the reader may be told the difference is exactly the same as in the case of water and steam. Heated in open vessels, water escapes as steam. Confined and heated in close vessels, it generates only the vapour requisite to develope a pressure sufficient to prevent further evaporation; the quantity of which vapour depends upon and varies with the temperature, increasing if heated higher, condensing if cooled. And precisely thus does carbonate of lime act, if heated in close vessels, giving off only a limited or definite quantity of *dissociated* carbonic acid gas, which varies in amount with the pressure and temperature, increasing if the mixture be made hotter, *recombining* chemically with the lime if cooled. Certainly, no single discovery in modern chemistry is more interesting than this.

A well-known difficulty, that of measuring very high temperatures, caused Debray to use those of boiling sulphur, cadmium, and zinc, which he estimated respectively at 440°, 860°, and 1040°. For the first of these, dissociation was inappreciable; for 880° it stopped when the tension became 85 millimètres, and for 1040° when the tension was 520 millimètres.

The difficulty just mentioned is readily overcome by substituting for water and carbonate of lime, first used by Deville and Debray, other and much more volatile and decomposable substances. This has been done by Jambert using for the purpose ammoniacal compounds. And the results obtained are of the most valuable and interesting nature. We regret that the scope of this article does not let us give a detailed account of them, but for that the reader may refer to the "Annales de l'École Normale," t. V, p. 129. His results were graphically represented by the method of curves, which are as exact and regular as they are for similar observations on steam.

200. Deville was the first to indicate how this subject should be mathematically studied and discussed. It is entirely unnecessary to go into that discussion, for it is sufficient to say the analogy between the dissociation of chemical compounds and the vaporization of a variable mixture of water and steam is perfect.

Every general formula is, therefore, at once applicable, and even every particular equation which we have deduced for mixtures of steam and water becomes immediately a formula for dissociation, if *mutatis mutandis* we substitute for steam the gases of decomposition, and for water the chemical compound.

It may be well to exemplify, and we select for the purpose, the law of Clapeyron

$$\lambda = A\tau\,(v - v_0)\,\frac{dp}{dt}.$$

In this λ now expresses the latent heat of expansion, or the energy requisite to do the work of chemical decomposition, we may call it the *latent heat of dissociation.* The volume of an unit weight of the compound is v_0, and v is that of its dissociated elements.

As λ will be positive, the last two factors of the second member are of like algebraic signs. When, therefore, v is greater than v_0, as is usually the case, the temperature of dissociation varies directly with the pressure; but one of these becomes a decreasing function of the other if v be less than v_0; as happens in the freezing of water. All this is evidently, by Art. 193, in such strict analogy that it is but repetition.

This formula of Clapeyron also expresses the heat developed in chemical combination; for taken inversely, let t and p both diminish, then will v also decrease; there will, therefore, be chemical reunion, or combination; and at the same time λ will decrease in quantity, or in other words heat will be set free or given out to surrounding bodies.

This conclusion is of the utmost value and importance, for it proves that we have in the formula of Clapeyron, or in the equivalent and corresponding formula of Sir W. Thomson, for the theorem of Carnot

$$l = A\tau\,\frac{dp}{dt},$$

a perfectly general law, applicable not only to all substances, but also to any and every physical or chemical change, whether of union or disunion of particles, in which heat is either absorbed or emitted, with variations of volume, pressure and temperature mutually dependent upon each other, so that they may be expressed by a determinate function

$$\phi\,(pvt) = 0\,;$$

whose precise form observation alone can and does determine for particular substances.

Moreover, the theorem of Clapeyron is one which lends itself with perfect facility to experimental investigation. Nothing is easier than to compute numerically the value of the derivative of the pressure from observed tables or data of tension; or graphically by curves, it may be found as the direction tangent of any elementary chord of the curve; while v and v_0 as well as λ are all most easily measured, requiring for this only skill, care and instrumental accuracy.

If heat be energy, what else is chemical "affinity"; and if they be mutually transformable into each other, and one is calculable, then assuredly the other is determinate; and we need only a factor like that of Joule to convert measurements of the one into their correlative values of the other. The only uncertainty about such calculations is that part of the energy may be dissipated in electrical effects, or lost as radiant energy in the form of luminiferous or other vibrations.

Here perhaps a protest may not be amiss against the practice of chemists who regard *affinity* as something *hyperphysical*,—something inscrutably mysterious, like the union of mind and matter, or like the spiritual part of our being. So long as it is thus regarded, there can be little hope of its ever being understood, even in matters perfectly accessible to investigation.

For the heat of chemical combination, or decomposition, or in other words for the value of λ, we already possess many experimental data, the most complete and exact of which are those of Favre and Silbermann. And all that is now needed, therefore, for applying the law of Clapeyron to such results of chemical combination is the experimental determination of the quantities v, v_0, p and τ contained in the second number.

201. The same formula also shows that λ is a function of and therefore varies with p, v and τ. And consequently, mere observations of λ, without regard to the circumstances upon which its variable values depend, are for exact science worthless. Unfortunately, we possess too many of such fictitious data, gathered with more laborious care and industry than intelligence.

ISOMERISM.

202. The transformation of paracyanogen by heat into cyanogen has been studied by MM. Troost and Hautefeuille. Heated under pressure p in a close vessel, the former a solid is changed into the latter a gas. Physically, this change is precisely analogous to that of water into steam, and the law of Clapeyron

$$\lambda = A\tau\,(v - v_0)\,\frac{dp}{dt}$$

is applicable.

We have now v for the specific volume of the mixture, or of the cyanogen, if v_0, which is that of the paracyanogen, be comparatively so small that it is negligible. And λ will be the heat or energy required for the transformation. As it is positive, and v greater than v_0, the pressure increases with the temperature. All of which agrees perfectly with the observed results.

The importance of these researches, which throw light upon what has hitherto been regarded as one of the most obscure subjects in chemistry, namely "isomerism" or "allotropism," is obvious, and they suggest many desirable investigations yet to be made.

DISSOCIATION OF CARBONIC OXIDE.

203. The discovery of Deville has already enriched experimental chemistry with many interesting facts; we will give only the following due to him.

A tube of brass runs concentrically through another of porcelain heated very highly. A rapid stream of cold water flows through the brass tube, and a current of carbonic oxide passes, very slowly, through the annular space between the brass and the porcelain tubes, into a solution of caustic potash. After some hours, the apparatus is dismounted, and the lower side of the brass tube is found coated with lamp black; also the solution of potash is found to contain carbonic acid.

Carbonic oxide has been dissociated. Its carbon has been deposited on the cold brass tube, and its oxygen has united with other portions of carbonic oxide to form carbonic acid. The experiment consists in chilling the carbon of the dissociated elements before it has recombined with oxygen, or reacted upon carbonic acid; effects which generally occur unless prevented.

The decomposing or reducing effect of solar rays upon the compounds of silver and other substances, used in photographic processes, will be recognized as phenomena closely analogous to, if not identical with, those of dissociation.

But by far the most important field of investigation in relation to dissociation is presented to our attention in the growth of plants; which take carbonic acid from the air, and decompose it by aid of solar energy into carbon and oxygen, storing up the first of these elements and emitting the other. Nor have we any right to suppose this decomposition an effect of light rather than of heat; for in fact radiant heat and light do not differ sufficiently to let them be assumed to be distinct and separate forms of physical energy.

STEADY FLOW OF FLUIDS.

204. In the steady flow of any fluid, let the pressure, specific volume, and height of a particle be respectively denoted by p, v, and z, and let u be its velocity; then will

$$\Sigma \int_1^0 P dp = \Sigma \frac{m}{2} (u^2 - u_0^2).$$

Or, supposing the unit of weight to be the quantity flowing through a section in a given time,

$$\Sigma \int_1^0 P dp = \frac{u^2 - u_0^2}{2g}.$$

To determine the first member, it divides itself into internal and external work, or

$$\Sigma \int_1^0 P dp = (U_0 - U) + (S_0 - S),$$

in which

$$(S_0 - S) = \int_1^0 p dv + (z_0 - z),$$

the last term denoting the work due to the fall of the unit weight from z_0 to z. Hence, by substitution,

$$\frac{u^2 - u_0^2}{2g} = (U_0 - U) + (p_0 v_0 - pv) + (z_0 - z), \qquad (201)$$

an equation perfectly general and applicable to all fluids whatever.

205. If the density be constant, the internal work will be nothing, and

$$\frac{u^2}{2g} + z + pv = \frac{u_0^2}{2g} + z_0 + p_0 v_0;$$

which is the *theorem of Bernouilli*. And if in this we suppose the fluid to start from rest, or u_0 equal zero, and that the pressure act like the atmosphere equally in both directions, then

$$u^2 = 2g (z_0 - z),$$

which is *Torricelli's law*.

When the fluid is a gas obeying the laws of Charles and Mariotte, or

$$pv = R\tau, \qquad \text{and} \qquad p_0 v_0 = R\tau_0,$$

equations (80) and (94) give

$$R = E(c' - c)$$

and

$$(U_0 - U) = Ec'(\tau_0 - \tau).$$

Whence, by substitution,

$$\frac{u^2 - u_0^2}{2g} = (z_0 - z) + Ec'(\tau_0 - \tau); \tag{202}$$

and if u_0 and the work $(z_0 - z)$ be negligible,

$$u^2 = 2Ec'g(\tau_0 - \tau). \tag{203}$$

206. For an adiabatic flow, or one in which time is not allowed for absorption or loss of heat, the equation

$$Q = A(U + S)$$

becomes

$$dU + pdv = 0;$$

whence

$$U_0 - U = \int_0^1 pdv,$$

or integrating by parts,

$$U_0 - U = pv - p_0 v_0 + \int_1^0 vdp.$$

Now, if this be substituted in equation (201) and z be negligible,

$$\frac{u_1^2 - u_0^2}{2g} = \int_1^0 vdp; \tag{204}$$

a very simple and general formula, in which v must be an adiabatic function of p, that it may be applicable.

207. To obtain an equation for the flow of saturated steam. We will suppose that a unit mixture of water and steam, or of moist vapour, moves rapidly from a position where its physical state is denoted by the co-ordinates p_0, v_0, and by the fraction of steam x_0, to another place where they become p, v, and x. Also that, as is usually the case, this movement and change of state takes place so quickly that heat cannot be absorbed or given off; or in other words, we assume the change to be adiabatic. Then clearly equation (177) furnishes the relation

$$\frac{\lambda_0 x_0}{\tau_0} = \frac{\lambda x}{\tau} + \int_{\tau_0}^{\tau} m' \frac{d\tau}{\tau}.$$

Also equation (183) gives

$$U - U_0 = E\left[\lambda x \frac{\tau - \tau_0}{\tau} + m'(\tau - \tau_0) + m'\tau_0 \log \frac{\tau_0}{\tau}\right]$$
$$- (pv - p_0 v_0) + \int_{p_0}^{p} s_0 dp.$$

Omitting z, this changes equation (201) into

$$\frac{u^2 - u_0^2}{2g} = s_0(p_0 - p) - E\left[\lambda x \frac{\tau - \tau_0}{\tau} + m'(\tau - \tau_0) + m'\tau_0 \log \frac{\tau_0}{\tau}\right].$$

A formula perfectly suited for numerical computation; and from which the data of Regnault easily show that in such changes the steam is partially liquefied.

CONCLUSION.

208. And now our task is finished. We have written what we undertook, not a complete treatise, but an elementary introduction to this important branch of physical science; and have done it in such a manner that he, who shall have faithfully studied what is here given, need not apprehend any difficulty in reading, with perfect ease, the original memoirs published upon this subject.

Much has been omitted; many varied and important applications to astronomy, to physical geography, and to electricity, are not even mentioned. It would require several volumes, not one alone, to present them to the attention of the reader. He need not be at a loss, however, to extend his knowledge, and we commend such inquiries to his consideration.

Again, whole branches of our knowledge concerning the phenomena and laws of heat are entirely excluded as forming no part of the special subject of this volume. Of such are Fourier's admirable investigations and their sequel. All too that relates to radiant heat.

To some it may seem strange, perhaps improper, that not one word is given about the theory of gases of Bernouilli, adopted by Clausius and others; which is dwelt upon in most treatises upon this subject, where the curious reader can readily find it. The omission is intentional. We have adhered rigidly and conscientiously to the purpose of excluding all hypothesis and speculation, and have presented nothing but what has been proved to be absolutely true.

It is only thus that positive truth can be separated from fiction, and presented as worthy of all confidence and acceptance. However ingenious, however suggestive of inquiry, an hypothesis may be, so long as it rests on mere unverified imagination, it must be discarded from that which is real and positive. Hypotheses have their legitimate use as means to ends, not as ends in themselves; they aid discovery, but are not discoveries; though in rare cases they sometimes become such, ceasing then to be hypotheses. Often they fetter rather than aid. And it is far easier to dream fiction, than it is with patient labour to discover and apply the laws of the universe whose Maker and Ruler is God.

It should always be borne in mind that false premises do by

logical reasoning lead to erroneous conclusions; and that only the true laws of the created world can be to us mental telescopes with which to penetrate its darkness.

In this day, when human perversity seems to incline some to prefer the arctic skepticism of negation to the genial warmth of Christian faith, while others go back to the atheistic blind necessity, the *ὑλικη ἀναγκη*, of Greek sophists, let it ever be remembered that, in the words of inspiration, "the invisible things of Him from the creation of the world are clearly seen, being understood by the things that are made."

May this book, whose object it is to make known some of those invisible things, aid its reader to form clearer ideas of the sublime simplicity, unity and harmony displayed by the Creator in the laws by which He governs the physical universe; some of which He does not conceal but graciously permits us to learn and understand.

May also the contemplation of the Infinite Wisdom, Power and Goodness manifested alike in the majestic laws and phenomena of the heavens, and in the infinitesimal adaptations of means to the support and happiness of every living creature, fit our hearts and minds for firmer and more grateful acceptance of that grandest of all truths—that He, "by whom were all things made, and without whom was not anything made that was made," did Himself become incarnate upon this earth, to the end that He might wipe out the imperfection of man's sinful and unredeemed nature, by Himself bearing "our sins in His body on the tree." And may human science learn that its highest duty is, on bended knee and with trembling but joyous heart, to point with uplifted finger steadfastly to the cross of Christ, man's only hope of a blessed immortality.

FINIS.

Francis' Lowell Hydraulics.

Third Edition.

4to. Cloth. $15.00.

LOWELL HYDRAULIC EXPERIMENTS — being a Selection from Experiments on Hydraulic Motors, on the Flow of Water over Weirs, and in Open Canals of Uniform Rectangular Section, made at Lowell, Mass. By J. B. FRANCIS, Civil Engineer. Third edition, revised and enlarged, including many New Experiments on Gauging Water in Open Canals, and on the Flow through Submerged Orifices and Diverging Tubes. With 23 copperplates, beautifully engraved, and about 100 new pages of text.

The work is divided into parts. PART I., on hydraulic motors, includes ninety-two experiments on an improved Fourneyron Turbine Water-Wheel, of about two hundred horse-power, with rules and tables for the construction of similar motors; thirteen experiments on a model of a centre-vent water-wheel of the most simple design, and thirty-nine experiments on a centre-vent water-wheel of about two hundred and thirty horse-power.

PART II. includes seventy-four experiments made for the purpose of determining the form of the formula for computing the flow of water over weirs; nine experiments on the effect of back-water on the flow over weirs; eighty-eight experiments made for the purpose of determining the formula for computing the flow over weirs of regular or standard forms, with several tables of comparisons of the new formula with the results obtained by former experimenters; five experiments on the flow over a dam in which the crest was of the same form as that built by the Essex Company across the Merrimack River at Lawrence, Massachusetts; twenty-one experiments on the effect of observing the depths of water on a weir at different distances from the weir; an extensive series of experiments made for the purpose of determining rules for gauging streams of water in open canals, with tables for facilitating the same; and one hundred and one experiments on the discharge of water through submerged orifices and diverging tubes, the whole being fully illustrated by twenty-three double plates engraved on copper.

In 1855 the proprietors of the Locks and Canals on Merrimack River consented to the publication of the first edition of this work, which contained a selection of the most important hydraulic experiments made at Lowell up to that time. In this edition the principal hydraulic experiments made there, subsequent to 1855, have been added, including the important series above mentioned, for determining rules for the gauging the flow of water in open canals, and the interesting series on the flow through a submerged Venturi's tube, in which a larger flow was obtained than any we find recorded.

Shreve on Bridges and Roofs.

8vo, 87 wood-cut illustrations. Cloth. $5.00.

A TREATISE ON THE STRENGTH OF BRIDGES AND ROOFS—comprising the determination of Algebraic formulas for Strains in Horizontal, Inclined or Rafter, Triangular, Bowstring, Lenticular and other Trusses, from fixed and moving loads, with practical applications and examples, for the use of Students and Engineers. By Samuel H. Shreve, A.M., Civil Engineer.

"On the whole, Mr. Shreve has produced a book which is the simplest, clearest, and at the same time, the most systematic and with the best mathematical reasoning of any work upon the same subject in the language."—*Railroad Gazette.*

"From the unusually clear language in which Mr. Shreve has given every statement, the student will have but himself to blame if he does not become thorough master of the subject."—*London Mining Journal.*

"Mr. Shreve has produced a work that must always take high rank as a text-book, * * * and no Bridge Engineer should be without it, as a valuable work of reference, and one that will frequently assist him out of difficulties."—*Franklin Institute Journal.*

The Kansas City Bridge.

4to. Cloth. $6.00

WITH AN ACCOUNT OF THE REGIMEN OF THE MISSOURI RIVER, and a description of the Methods used for Founding in that River. By O. Chanute, Chief Engineer, and George Morison, Assistant Engineer. Illustrated with five lithographic views and twelve plates of plans.

Illustrations.

Views.—View of the Kansas City Bridge, August 2, 1869. Lowering Caisson No. 1 into position. Caisson for Pier No. 4 brought into position. View of Foundation Works, Pier No. 4. Pier No. 1.

Plates.—I. Map showing location of Bridge. II. Water Record—Cross Section of River—Profile of Crossing—Pontoon Protection. III. Water Deadener—Caisson No. 2—Foundation Works, Pier No. 3. IV. Foundation Works, Pier No. 4. V. Foundation Works, Pier No. 4. VI. Caisson No. 5—Sheet Piling at Pier No. 6—Details of Dredges—Pile Shoe—Beton Box. VII. Masonry—Draw Protection—False Works between Piers 3 and 4. VIII. Floating Derricks. IX. General Elevation—176 feet span. X. 248 feet span. XI. Plans of Draw. XII. Strain Diagrams.

Clarke's Quincy Bridge.

4to. Cloth. $7.50.

DESCRIPTION OF THE IRON RAILWAY Bridge across the Mississippi River at Quincy, Illinois. By THOMAS CURTIS CLARKE, Chief Engineer. Illustrated with twenty-one lithographed plans.

Barba on the Use of Steel.

12mo. Illustrated. Cloth. $1.50.

THE USE OF STEEL IN CONSTRUCTION. Method of Working, Applying, and Testing Plates and Bars. By J. BARBA, Chief Naval Constructor. Translated from the French, with a Preface, by A. L. HOLLEY, P.B.

Whipple on Bridge Building.

8vo, Illustrated. Cloth. $4.00.

AN ELEMENTARY AND PRACTICAL TREATISE ON BRIDGE BUILDING. An enlarged and improved edition of the Author's original work. By S. WHIPPLE, C. E., Inventor of the Whipple Bridges, &c. **Second Edition.**

The design has been to develop from Fundamental Principles a system easy of comprehension, and such as to enable the attentive reader and student to judge understandingly for himself, as to the relative merits of different plans and combinations, and to adopt for use such as may be most suitable for the cases he may have to deal with.

It is hoped the work may prove an appropriate Text-Book upon the subject treated of, for the Engineering Student, and a useful manual for the Practicing Engineer and Bridge Builder.

Stoney on Strains.

New and Revised Edition, with numerous illustrations.

Royal 8vo, 664 pp. Cloth. $12.50.

THE THEORY OF STRAINS IN GIRDERS and Similar Structures, with Observations on the Application of Theory to Practice, and Tables of Strength and other Properties of Materials. By BINDON B. STONEY, B. A.

Roebling's Bridges.

Imperial folio. Cloth. $25.00.

LONG AND SHORT SPAN RAILWAY BRIDGES. By JOHN A. ROEBLING, C. E. Illustrated with large copperplate engravings of plans and views.

List of Plates

1. Parabolic Truss Railway Bridge. 2, 3, 4, 5, 6. Details of Parabolic Truss, with centre span 500 feet in the clear. 7. Plan and View of a Bridge over the Mississippi River, at St. Louis, for railway and common travel. 8, 9, 10, 11, 12. Details and View of St. Louis Bridge. 13. Railroad Bridge over the Ohio.

Diedrichs' Theory of Strains.

8vo. Cloth. $5.00.

A Compendium for the Calculation and Construction of Bridges, Roofs, and Cranes, with the Application of Trigonometrical Notes. Containing the most comprehensive information in regard to the Resulting Strains for a permanent Load, as also for a combined (Permanent and Rolling) Load. In two sections adapted to the requirements of the present time. By JOHN DIEDRICHS. Illustrated by numerous plates and diagrams.

"The want of a compact, universal and popular treatise on the Construction of Roofs and Bridges—especially one treating of the influence of a variable load—and the unsatisfactory essays of different authors on the subject, induced me to prepare this work."

Jacob on Retaining Walls.

18mo. Boards. 50 cts.

PRACTICAL DESIGNING OF RETAINING WALLS. By ARTHUR JACOB, A. B.

Campin on Iron Roofs.

Large 8vo. Cloth. $2.00.

ON THE CONSTRUCTION OF IRON ROOFS. A Theoretical and Practical Treatise. By FRANCIS CAMPIN. With wood-cuts and plates of Roofs lately executed.

"The mathematical formulas are of an elementary kind, and the process admits of an easy extension so as to embrace the prominent varieties of iron truss bridges. The treatise, though of a practical scientific character, may be easily mastered by any one familiar with elementary mechanics and plane trigonometry."

Holley's Railway Practice.

1 vol. folio. Cloth. $12.00.

AMERICAN AND EUROPEAN RAILWAY PRACTICE, in the Economical Generation of Steam, including the materials and construction of Coal-burning Boilers, Combustion, the Variable Blast, Vaporization, Circulation, Super-heating, Supplying and Heating Feed-water, &c., and the adaptation of Wood and Coke-burning Engines to Coal-burning; and in Permanent Way, including Road-bed, Sleepers, Rails, Joint Fastenings, Street Railways, &c., &c. By ALEXANDER L. HOLLEY, B. P. With 77 lithographed plates.

"This is an elaborate treatise by one of our ablest civil engineers, on the construction and use of locomotives, with a few chapters on the building of Railroads. * * * All these subjects are treated by the author, who is a first-class railroad engineer, in both an intelligent and intelligible manner. The facts and ideas are well arranged, and presented in a clear and simple style, accompanied by beautiful engravings, and we presume the work will be regarded as indispensable by all who are interested in a knowledge of the construction of railroads and rolling stock, or the working of locomotives."—*Scientific American.*

Henrici's Skeleton Structures.

8vo. Cloth. $1.50.

SKELETON STRUCTURES, especially in their Application to the building of Steel and Iron Bridges. By Olaus Henrici. With folding plates and diagrams.

By presenting these general examinations on Skeleton Structures, with particular application for Suspended Bridges, to Engineers, I venture to express the hope that they will receive these theoretical results with some confidence, even although an opportunity is wanting to compare them with practical results. O. H.

Useful Information for Railway Men.

Pocket form. Morocco, gilt, $2.00.

Compiled by W. G. Hamilton, Engineer. Sixth edition, revised and enlarged. 570 pages.

"It embodies many valuable formulæ and recipes useful for railway men, and, indeed, for almost every class of persons in the world. The 'information' comprises some valuable formulæ and rules for the construction of boilers and engines, masonry, properties of steel and iron, and the strength of materials generally."—*Railroad Gazette, Chicago.*

The Mechanic's Friend.

12mo. Cloth. 300 Illustrations. $1.50.

THE MECHANIC'S FRIEND: A Collection of Receipts and Practical Suggestions, Relating to Aquaria — Bronzing — Cements — Drawing — Dyes — Electricity — Gilding — Glass-Working — Glues — Horology — Lacquers — Locomotives — Magnetism — Metal-Working — Modelling — Photography — Pyrotechny — Railways. — Solders — Steam-Engine — Telegraphy — Taxidermy — Varnishes — Waterproofing — and Miscellaneous Tools, Instruments, Machines, and Processes connected with the Chemical and Mechanical Arts. By William E. Axon, M.R.S.L.

Kirkwood on Filtration.

4to. Cloth. $15.00.

REPORT ON THE FILTRATION OF RIVER WATERS, for the Supply of Cities, as practised in Europe, made to the Board of Water Commissioners of the City of St. Louis. By JAMES P. KIRKWOOD. Illustrated by 30 double-plate engravings.

CONTENTS.—Report on Filtration—London Works, General—Chelsea Water Works and Filters—Lambeth Water Works and Filters—Southwark and Vauxhall Water Works and Filters—Grand Junction Water Works and Filters—West Middlesex Water Works and Filters—New River Water Works and Filters—East London Water Works and Filters—Leicester Water Works and Filters—York Water Works and Filters—Liverpool Water Works and Filters—Edinburgh Water Works and Filters—Dublin Water Works and Filters—Perth Water Works and Filtering Gallery—Berlin Water Works and Filters—Hamburg Water Works and Reservoirs—Altona Water Works and Filters—Tours Water Works and Filtering Canal—Angers Water Works and Filtering Galleries—Nantes Water Works and Filters—Lyons Water Works and Filtering Galleries—Toulouse Water Works and Filtering Galleries—Marseilles Water Works and Filters—Genoa Water Works and Filtering Galleries—Leghorn Water Works and Cisterns—Wakefield Water Works and Filters—Appendix.

Tunner on Roll-Turning.

1 vol. 8vo. and 1 vol. plates. $10.00.

A TREATISE ON ROLL-TURNING FOR THE MANUFACTURE OF IRON. By PETER TUNNER. Translated and adapted. By JOHN B. PEARSE, of the Pennsylvania Steel Works. With numerous wood-cuts, 8vo., together with a folio atlas of 10 lithographed plates of Rolls, Measurements, &c.

"We commend this book as a clear, elaborate, and practical treatise upon the department of iron manufacturing operations to which it is devoted. The writer states in his preface, that for twenty-five years he has felt the necessity of such a work, and has evidently brought to its preparation the fruits of experience, a painstaking regard for accuracy of statement, and a desire to furnish information in a style readily understood. The book should be in the hands of every one interested, either in the general practice of mechanical engineering, or the special branch of manufacturing operations to which the work relates.'—*American Artisan.*

Jacob on Storage Reservoirs.

18mo. Boards. 50 cts.

THE DESIGNING AND CONSTRUCTION OF STORAGE RESERVOIRS. By ARTHUR JACOB, B. A. With tables and wood-cuts representing sections, etc.

Hewson on Embankments.

8vo. Cloth. $2.00.

PRINCIPLES AND PRACTICE OF EMBANKING LANDS from River Floods, as applied to the Levees of the Mississippi. By WILLIAM HEWSON, Civil Engineer.

"This is a valuable treatise on the principles and practice of embanking lands from river floods, as applied to the Levees of the Mississippi, by a highly intelligent and experienced engineer. The author says it is a first attempt to reduce to order and to rule the design, execution, and measurement of the Levees of the Mississippi. It is a most useful and needed contribution to scientific literature.—*Philadelphia Evening Journal.*

Grüner on Steel.

8vo. Cloth. $3.50.

THE MANUFACTURE OF STEEL. By M. L. GRUNER, translated from the French. By Lenox Smith, A. M., E. M., with an appendix on the Bessemer Process in the United States, by the translator. Illustrated by lithographed drawings and wood-cuts.

"The purpose of the work is to present a careful, elaborate, and at the same time practical examination into the physical properties of steel, as well as a description of the new processes and mechanical appliances for its manufacture. The information which it contains, gathered from many trustworthy sources, will be found of much value to the American steel manufacturer, who may thus acquaint himself with the results of careful and elaborate experiments in other countries, and better prepare himself for successful competition in this important industry with foreign makers. The fact that this volume is from the pen of one of the ablest metallurgists of the present day, cannot fail, we think, to secure for it a favorable consideration.—*Iron Age.*

Bauerman on Iron.

12mo. Cloth. $2.00.

TREATISE ON THE METALLURGY OF IRON. Containing outlines of the History of Iron Manufacture, methods of Assay, and analysis of Iron Ores, processes of manufacture of Iron and Steel, etc., etc. By H. Bauerman. First American edition. Revised and enlarged, with an appendix on the Martin Process for making Steel, from the report of Abram S. Hewitt. Illustrated with numerous wood engravings.

"This is an important addition to the stock of technical works published in this country. It embodies the latest facts, discoveries, and processes connected with the manufacture of iron and steel, and should be in the hands of every person interested in the subject, as well as in all technical and scientific libraries."—*Scientific American.*

Link and Valve Motions, by W. S. Auchincloss.

Sixth Edition. 8vo. Cloth. $3.00.

APPLICATION OF THE SLIDE VALVE and Link Motion to Stationary, Portable, Locomotive and Marine Engines, with new and simple methods for proportioning the parts. By William S. Auchincloss, Civil and Mechanical Engineer. Designed as a hand-book for Mechanical Engineers, Master Mechanics, Draughtsmen and Students of Steam Engineering. All dimensions of the valve are found with the greatest ease by means of a Printed Scale, and proportions of the link determined *without* the assistance of a model. Illustrated by 37 wood-cuts and 21 lithographic plates, together with a copperplate engraving of the Travel Scale.

All the matters we have mentioned are treated with a clearness and absence of unnecessary verbiage which renders the work a peculiarly valuable one. The Travel Scale only requires to be known to be appreciated. Mr. A. writes so ably on his subject, we wish he had written more. *London Engineering.*

We have never opened a work relating to steam which seemed to us better calculated to give an intelligent mind a clear understanding of the department it discusses.—*Scientific American.*

Slide Valve by Eccentrics, by Prof. C. W. MacCord.

4to. Illustrated. Cloth, $3.00.

A PRACTICAL TREATISE ON THE SLIDE VALVE BY ECCENTRICS, examining by methods, the action of the Eccentric upon the Slide Valve, and explaining the practical processes of laying out the movements, adapting the valve for its various duties in the steam-engine. For the use of Engineers, Draughtsmen, Machinists, and Students of valve motions in general. By C. W. MacCord, A. M., Professor of Mechanical Drawing, Stevens' Institute of Technology, Hoboken, N J.

Stillman's Steam-Engine Indicator.

12mo. Cloth. $1.00.

THE STEAM-ENGINE INDICATOR, and the Improved Manometer Steam and Vacuum Gauges; their utility and application By Paul Stillman. New edition.

Bacon's Steam-Engine Indicator.

12mo. Cloth. $1.00. Mor. $1.50.

A TREATISE ON THE RICHARDS STEAM-ENGINE INDICATOR, with directions for its use. By Charles T. Porter. Revised, with notes and large additions as developed by American Practice, with an Appendix containing useful formulæ and rules for Engineers. By F. W. Bacon, M. E., Member of the American Society of Civil Engineers. Illustrated. **Second Edition**

In this work, Mr. Porter's book has been taken as the basis, but Mr. Bacon has adapted it to American Practice, and has conferred a great boon on American Engineers.—*Artisan.*

Gillmore's Building Stones.

8vo. Cloth. $1 50.

REPORT ON STRENGTH OF THE BUILDING STONES IN THE UNITED STATES, Etc.

Gillmore's Limes and Cements.

Fifth Edition. Revised and Enlarged.

8vo. Cloth. $4.00.

PRACTICAL TREATISE ON LIMES, HYDRAULIC CEMENTS, AND MORTARS. Papers on Practical Engineering, U. S. Engineer Department, No. 9, containing Reports of numerous experiments conducted in New York City, during the years 1858 to 1861, inclusive. By Q. A. GILLMORE, Lt.-Col. U. S. Corps of Engineers, Brevet Major-General U. S. Army. With numerous illustrations.

"This work contains a record of certain experiments and researches made under the authority of the Engineer Bureau of the War Department from 1858 to 1861, upon the various hydraulic cements of the United States, and the materials for their manufacture. The experiments were carefully made, and are well reported and compiled.'—*Journal Franklin Institute.*

Gillmore's Coignet Beton.

8vo. Cloth. $2.50.

COIGNET BETON AND OTHER ARTIFICIAL STONE. By Q. A. GILLMORE, Lt.-Col. U. S Corps of Engineers, Brevet Major-General U. S. Army. 9 Plates, Views, etc.

This work describes with considerable minuteness of detail the several kinds of artificial stone in most general use in Europe and now beginning to be introduced in the United States, discusses their properties, relative merits, and cost, and describes the materials of which they are composed. . . . . The subject is one of special and growing interest, and we commend the work, embodying as it does the matured opinions of an experienced engineer and expert.

Gillmore on Roads.

12mo. Cloth. $2.00.

A PRACTICAL TREATISE ON THE CONSTRUCTION OF ROADS, STREETS, AND PAVEMENTS. By Q. A. GILLMORE, Lt.-Col. U. S. Corps of Engineers, Brevet Major-General U. S. Army.

Williamson on the Barometer.

4to. Cloth. $15.00.

ON THE USE OF THE BAROMETER ON SURVEYS AND RECONNAISSANCES. Part I. Meteorology in its Connection with Hypsometry. Part II. Barometric Hypsometry. By R. S. Williamson, Bvt. Lieut.-Col. U. S. A., Major Corps of Engineers. With Illustrative Tables and Engravings. Paper No. 15, Professional Papers, Corps of Engineers.

"San Francisco, Cal., *Feb.* 27, 1867.

"Gen. A. A. Humphreys, Chief of Engineers, U. S. Army:

"General,—I have the honor to submit to you, in the following pages, the results of my investigations in meteorology and hypsometry, made with the view of ascertaining how far the barometer can be used as a reliable instrument for determining altitudes on extended lines of survey and reconnaissances. These investigations have occupied the leisure permitted me from my professional duties during the last ten years, and I hope the results will be deemed of sufficient value to have a place assigned them among the printed professional papers of the United States Corps of Engineers.

"Very respectfully, your obedient servant,

"R. S. WILLIAMSON,

"Bvt. Lt.-Col. U. S. A., Major Corps of U. S. Engineers."

Von Cotta's Ore Deposits.

8vo. Cloth. $4.00.

TREATISE ON ORE DEPOSITS. By Bernhard Von Cotta, Professor of Geology in the Royal School of Mines, Freidberg, Saxony. Translated from the second German edition, by Frederick Prime, Jr., Mining Engineer, and revised by the author, with numerous illustrations.

"Prof. Von Cotta of the Freiberg School of Mines, is the author of the best modern treatise on ore deposits, and we are heartily glad that this admirable work has been translated and published in this country. The translator, Mr. Frederick Prime, Jr., a graduate of Freiberg, has had in his work the great advantage of a revision by the author himself, who declares in a prefatory note that this may be considered as a new edition (the third) of his own book.

"It is a timely and welcome contribution to the literature of mining in this country, and we are grateful to the translator for his enterprise and good judgment in undertaking its preparation; while we recognize with equal cordiality the liberality of the author in granting both permission and assistance."—*Extract from Review in Engineering and Mining Journal.*

Plattner's Blow-Pipe Analysis.

Second edition. Revised. 8vo. Cloth. $7.50.

PLATTNER'S MANUAL OF QUALITATIVE AND QUANTITATIVE ANALYSIS WITH THE BLOW-PIPE. From the last German edition Revised and enlarged. By Prof. TH. RICHTER, of the Royal Saxon Mining Academy. Translated by Prof. H. B. CORNWALL, Assistant in the Columbia School of Mines, New York; assisted by JOHN H. CASWELL. Illustrated with eighty-seven wood-cuts and one Lithographic Plate. 560 pages.

"Plattner's celebrated work has long been recognized as the only complete book on Blow-Pipe Analysis. The fourth German edition, edited by Prof. Richter, fully sustains the reputation which the earlier editions acquired during the lifetime of the author, and it is a source of great satisfaction to us to know that Prof. Richter has co-operated with the translator in issuing the American edition of the work, which is in fact a fifth edition of the original work, being far more complete than the last German edition."—*Silliman's Journal.*

There is nothing so complete to be found in the English language. Plattner's book is not a mere pocket edition; it is intended as a comprehensive guide to all that is at present known on the blow-pipe, and as such is really indispensable to teachers and advanced pupils.

"Mr. Cornwall's edition is something more than a translation, as it contains many corrections, emendations and additions not to be found in the original. It is a decided improvement on the work in its German dress."—*Journal of Applied Chemistry.*

Egleston's Mineralogy.

8vo. Illustrated with 34 Lithographic Plates. Cloth. $4.50.

LECTURES ON DESCRIPTIVE MINERALOGY, Delivered at the School of Mines, Columbia College. BY PROFESSOR T. EGLESTON.

These lectures are what their title indicates, the lectures on Mineralogy delivered at the School of Mines of Columbia College. They have been printed for the students, in order that more time might be given to the various methods of examining and determining minerals. The second part has only been printed. The first part, comprising crystallography and physical mineralogy, will be printed at some future time.

Pynchon's Chemical Physics.

New Edition. Revised and Enlarged.

Crown 8vo. Cloth. $3.00.

INTRODUCTION TO CHEMICAL PHYSICS, Designed for the Use of Academies, Colleges, and High Schools. Illustrated with numerous engravings, and containing copious experiments with directions for preparing them. By THOMAS RUGGLES PYNCHON, M.A., Professor of Chemistry and the Natural Sciences, Trinity College, Hartford.

Hitherto, no work suitable for general use, treating of all these subjects within the limits of a single volume, could be found; consequently the attention they have received has not been at all proportionate to their importance. It is believed that a book containing so much valuable information within so small a compass, cannot fail to meet with a ready sale among all intelligent persons, while Professional men, Physicians, Medical Students, Photographers, Telegraphers, Engineers, and Artisans generally, will find it specially valuable, if not nearly indispensable, as a book of reference.

"We strongly recommend this able treatise to our readers as the first work ever published on the subject free from perplexing technicalities. In style it is pure, in description graphic, and its typographical appearance is artistic. It is altogether a most excellent work."—*Eclectic Medical Journal.*

"It treats fully of Photography, Telegraphy, Steam Engines, and the various applications of Electricity. In short, it is a carefully prepared volume, abreast with the latest scientific discoveries and inventions."—*Hartford Courant.*

Plympton's Blow-Pipe Analysis.

12mo. Cloth. $1 50.

THE BLOW-PIPE: A Guide to Its Use in the Determination of Salts and Minerals. Compiled from various sources, by GEORGE W. PLYMPTON, C.E., A.M., Professor of Physical Science in the Polytechnic Institute, Brooklyn, N. Y.

"This manual probably has no superior in the English language as a text-book for beginners, or as a guide to the student working without a teacher. To the latter many illustrations of the utensils and apparatus required in using the blow-pipe, as well as the fully illustrated description of the blow-pipe flame, will be especially serviceable."—*New York Teacher.*

Dubois' Graphical Statics.

8vo. 60 Illustrations. Cloth. $2.00.

THE NEW METHOD OF GRAPHICAL STATICS. By A. J. DUBOIS, C.E., Ph.D.

Gases in Coal Mines

18mo. Boards. 50 cts.

A PRACTICAL TREATISE ON THE GASES MET WITH IN COAL MINES. By the late J. J. ATKINSON, Government Inspector of Mines for the County of Durham, England.

Watt's Dictionary of Chemistry.

Supplementary Volume.

8vo. Cloth. $9.00.

This volume brings the Record of Chemical Discovery down to the end of the year 1869, including also several additions to, and corrections of, former results which have appeared in 1870 and 1871.

*** Complete Sets of the Work, New and Revised edition, including above supplement. 6 vols. 8vo. Cloth. $62.00.

Rammelsberg's Chemical Analysis.

8vo. Cloth. $2.25.

GUIDE TO A COURSE OF QUANTITATIVE CHEMICAL ANALYSIS, ESPECIALLY OF MINERALS AND FURNACE PRODUCTS. Illustrated by Examples. By C. F. RAMMELSBERG. Translated by J. TOWLER, M.D.

This work has been translated, and is now published expressly for those students in chemistry whose time and other studies in colleges do not permit them to enter upon the more elaborate and expensive treatises of Fresenius and others. It is the condensed labor of a master in chemistry and of a practical analyst.

Eliot and Storer's Qualitative Chemical Analysis.

New Edition, Revised.

12mo. Illustrated. Cloth. $1.50.

A COMPENDIOUS MANUAL OF QUALITATIVE CHEMICAL ANALYSIS. By Charles W. Eliot and Frank H. Storer. Revised with the Coöperation of the Authors, by William Ripley Nichols, Professor of Chemistry in the Massachusetts Institute of Technology.

"This Manual has great merits as a practical introduction to the science and the art of which it treats. It contains enough of the theory and practice of qualitative analysis, "in the wet way," to bring out all the reasoning involved in the science, and to present clearly to the student the most approved methods of the art. It is specially adapted for exercises and experiments in the laboratory; and yet its classifications and manner of treatment are so systematic and logical throughout, as to adapt it in a high degree to that higher class of students generally who desire an accurate knowledge of the practical methods of arriving at scientific facts."—*Lutheran Observer.*

"We wish every academical class in the land could have the benefit of the fifty exercises of two hours each necessary to master this book. Chemistry would cease to be a mere matter of memory, and become a pleasant experimental and intellectual recreation. We heartily commend this little volume to the notice of those teachers who believe in using the sciences as means of mental discipline."—*College Courant.*

Craig's Decimal System.

Square 32mo. Limp. 50c.

WEIGHTS AND MEASURES. An Account of the Decimal System, with Tables of Conversion for Commercial and Scientific Uses. By B. F. Craig, M. D.

"The most lucid, accurate, and useful of all the hand-books on this subject that we have yet seen. It gives forty-seven tables of comparison between the English and French denominations of length, area, capacity, weight, and the Centigrade and Fahrenheit thermometers, with clear instructions how to use them; and to this practical portion, which helps to make the transition as easy as possible, is prefixed a scientific explanation of the errors in the metric system, and how they may be corrected in the laboratory."—*Nation.*

Nugent on Optics.

12mo. Cloth. $2.00

TREATISE ON OPTICS; or, Light and Sight, theoretically and practically treated; with the application to Fine Art and Industrial Pursuits. By E. NUGENT. With one hundred and three illustrations.

"This book is of a practical rather than a theoretical kind, and is designed to afford accurate and complete information to all interested in applications of the science."—*Round Table.*

Barnard's Metric System.

8vo. Brown cloth. $3.00.

THE METRIC SYSTEM OF WEIGHTS AND MEASURES. An Address delivered before the Convocation of the University of the State of New York, at Albany, August, 1871. By FREDERICK A. P. BARNARD, President of Columbia College, New York City. Second edition from the Revised edition printed for the Trustees of Columbia College. Tinted paper.

"It is the best summary of the arguments in favor of the metric weights and measures with which we are acquainted, not only because it contains in small space the leading facts of the case, but because it puts the advocacy of that system on the only tenable grounds, namely, the great convenience of a decimal notation of weight and measure as well as money, the value of international uniformity in the matter, and the fact that this metric system is adopted and in general use by the majority of civilized nations."—*The Nation.*

Butler on Ventilation.

18mo. Boards. 50 cts.

VENTILATION OF BUILDINGS. By W. F. BUTLER. Illustrated.

"As death by insensible suffocation is one of the prominent causes which swell our bills of mortality, we commend this book to the attention of philanthropists as well as to architects."—*Boston Globe.*

Harrison's Mechanic's Tool-Book.

12mo. Cloth. $1.50.

MECHANIC'S TOOL BOOK, with practical rules and suggestions, for the use of Machinists, Iron Workers, and others. By W. B. Harrison, Associate Editor of the "American Artisan." Illustrated with 44 engravings.

"This work is specially adapted to meet the wants of Machinists and workers in iron generally. It is made up of the work-day experience of an intelligent and ingenious mechanic, who had the faculty of adapting tools to various purposes. The practicability of his plans and suggestions are made apparent even to the unpractised eye by a series of well-executed wood engravings."—*Philadelphia Inquirer.*

Pope's Modern Practice of the Electric Telegraph.

Ninth Edition. 8vo. Cloth $2.00.

A Hand-book for Electricians and Operators. By Frank L. Pope. Seventh edition. Revised and enlarged, and fully illustrated.

Extract from Letter of Prof. Morse.

"I have had time only cursorily to examine its contents, but this examination has resulted in great gratification, especially at the fairness and unprejudiced tone of your whole work.

"Your illustrated diagrams are admirable and beautifully executed.

"I think all your instructions in the use of the telegraph apparatus judicious and correct, and I most cordially wish you success."

Extract from Letter of Prof. G. W. Hough, of the Dudley Observatory.

"There is no other work of this kind in the English language that contains in so small a compass so much practical information in the application of galvanic electricity to telegraphy. It should be in the hands of every one interested in telegraphy, or the use of Batteries for other purposes."

Morse's Telegraphic Apparatus.

Illustrated. 8vo. Cloth. $2.00.

EXAMINATION OF THE TELEGRAPHIC APPARATUS AND THE PROCESSES IN TELEGAPHY. By Samuel F. B. Morse, LL.D., United States Commissioner Paris Universal Exposition, 1867.

Sabine's History of the Telegraph.

12mo. Cloth. $1.25.

HISTORY AND PROGRESS OF THE ELECTRIC TELEGRAPH, with Descriptions of some of the Apparatus. By Robert Sabine, C. E. Second edition, with additions.

Contents.—I. Early Observations of Electrical Phenomena. II. Telegraphs by Frictional Electricity. III. Telegraphs by Voltaic Electricity. IV. Telegraphs by Electro-Magnetism and Magneto-Electricity. V. Telegraphs now in use. VI. Overhead Lines. VII. Submarine Telegraph Lines. VIII. Underground Telegraphs. IX. Atmospheric Electricity.

Haskins' Galvanometer.

Pocket form. Illustrated. Morocco tucks. $2.00.

THE GALVANOMETER, AND ITS USES; a Manual for Electricians and Students. By C. H. Haskins.

"We hope this excellent little work will meet with the sale its merits entitle it to. To every telegrapher who owns, or uses a Galvanometer, or ever expects to, it will be quite indispensable."—*The Telegrapher.*

Culley's Hand-Book of Telegraphy.

8vo. Cloth. $5.00.

A HAND-BOOK OF PRACTICAL TELEGRAPHY. By R. S. Culley, Engineer to the Electric and International Telegraph Company. Fifth edition, revised and enlarged.

Foster's Submarine Blasting.

4to. Cloth. $3.50.

SUBMARINE BLASTING in Boston Harbor, Massachusetts—Removal of Tower and Corwin Rocks. By John G. Foster, Lieutenant-Colonel of Engineers, and Brevet Major-General, U. S. Army. Illustrated with seven plates.

List of Plates.—1. Sketch of the Narrows, Boston Harbor. 2. Townsend's Submarine Drilling Machine, and Working Vessel attending. 3. Submarine Drilling Machine employed. 4. Details of Drilling Machine employed. 5. Cartridges and Tamping used. 6. Fuses and Insulated Wires used. 7. Portable Friction Battery used.

Barnes' Submarine Warfare.

8vo. Cloth. $5.00.

SUBMARINE WARFARE, DEFENSIVE AND OFFENSIVE. Comprising a full and complete History of the Invention of the Torpedo, its employment in War and results of its use. Descriptions of the various forms of Torpedoes, Submarine Batteries and Torpedo Boats actually used in War. Methods of Ignition by Machinery, Contact Fuzes, and Electricity, and a full account of experiments made to determine the Explosive Force of Gunpowder under Water. Also a discussion of the Offensive Torpedo system, its effect upon Iron-Clad Ship systems, and influence upon Future Naval Wars. By Lieut.-Commander JOHN S. BARNES, U. S. N. With twenty lithographic plates and many wood-cuts.

"A book important to military men, and especially so to engineers and artillerists. It consists of an examination of the various offensive and defensive engines that have been contrived for submarine hostilities, including a discussion of the torpedo system, its effects upon iron-clad ship-systems, and its probable influence upon future naval wars. Plates of a valuable character accompany the treatise, which affords a useful history of the momentous subject it discusses. A great deal of useful information is collected in its pages, especially concerning the inventions of SCHOLL and VERDU, and of JONES' and HUNT'S batteries, as well as of other similar machines, and the use in submarine operations of gun-cotton and nitro-glycerine."—*N. Y. Times.*

Randall's Quartz Operator's Hand-Book.

12mo. Cloth. $2.00.

QUARTZ OPERATOR'S HAND-BOOK. By P. M. RANDALL. New edition, revised and enlarged. Fully illustrated.

The object of this work has been to present a clear and comprehensive exposition of mineral veins, and the means and modes chiefly employed for the mining and working of their ores—more especially those containing gold and silver.

McCulloch's Theory of Heat.

8vo. Cloth. In Press.

AN ELEMENTARY TREATISE ON THE MECHANICAL THEORY OF HEAT, AND ITS APPLICATION TO AIR AND STEAM ENGINES. By Prof. R. S. McCulloch.

Benét's Chronoscope.

Second Edition.

Illustrated. 4to. Cloth. $3.00.

ELECTRO-BALLISTIC MACHINES, and the Schultz Chronoscope. By Lieutenant-Colonel S. V. Benét, Captain of Ordnance, U. S. Army.

Contents.—1. Ballistic Pendulum. 2. Gun Pendulum. 3. Use of Electricity. 4. Navez' Machine. 5. Vignotti's Machine, with Plates. 6. Benton's Electro-Ballistic Pendulum, with Plates. 7. Leur's Tro-Pendulum Machine 8. Schultz's Chrcnoscope, with two Plates.

Michaelis' Chronograph.

4to. Illustrated. Cloth. $3.00.

THE LE BOULENGÉ CHRONOGRAPH. With three lithographed folding plates of illustrations. By Brevet Captain O E. Michaelis, First Lieutenant Ordnance Corps, U. S. Army.

"The excellent monograph of Captain Michaelis enters minutely into the details of construction and management, and gives tables of the times of flight calculated upon a given fall of the chronometer for all distances. Captain Michaelis has done good service in presenting this work to his brother officers, describing, as it does, an instrument which bids fair to be in constant use in our future ballistic experiments."—*Army and Navy Journal.*

Silversmith's Hand-Book.

Fourth Edition.

Illustrated. 12mo. Cloth. $3.00.

A PRACTICAL HAND-BOOK FOR MINERS, Metallurgists, and Assayers, comprising the most recent improvements in the disintegration, amalgamation, smelting, and parting of the Precious Ores, with a Comprehensive Digest of the Mining Laws. Greatly augmented, revised, and corrected. By JULIUS SILVERSMITH. Fourth edition. Profusely illustrated. 1 vol. 12mo. Cloth. $3.00.

One of the most important features of this work is that in which the metallurgy of the precious metals is treated of. In it the author has endeavored to embody all the processes for the reduction and manipulation of the precious ores heretofore successfully employed in Germany, England, Mexico, and the United States, together with such as have been more recently invented, and not yet fully tested—all of which are profusely illustrated and easy of comprehension.

Simms' Levelling.

8vo. Cloth. $2.50.

A TREATISE ON THE PRINCIPLES AND PRACTICE OF LEVELLING, showing its application to purposes of Railway Engineering and the Construction of Roads, &c. By FREDERICK W. SIMMS, C. E. From the fifth London edition, revised and corrected, with the addition of Mr. Law's Practical Examples for Setting Out Railway Curves. Illustrated with three lithographic plates and numerous wood-cuts.

"One of the most important text-books for the general surveyor, and there is scarcely a question connected with levelling for which a solution would be sought, but that would be satisfactorily answered by consulting this volume." —*Mining Journal.*

"The text-book on levelling in most of our engineering schools and colleges."—*Engineers.*

"The publishers have rendered a substantial service to the profession, especially to the younger members, by bringing out the present edition of Mr. Simms' useful work."—*Engineering.*

Stuart's Successful Engineer.

18mo. Boards. 50 cents.

HOW TO BECOME A SUCCESSFUL ENGINEER: Being Hints to Youths intending to adopt the Profession. By BERNARD STUART, Engineer. Sixth Edition.

"A valuable little book of sound, sensible advice to young men who wish to rise in the most important of the professions."—*Scientific American.*

Stuart's Naval Dry Docks.

Twenty-four engravings on steel.

Fourth Edition.

4to. Cloth. $6.00.

THE NAVAL DRY DOCKS OF THE UNITED STATES. By CHARLES B. STUART. Engineer in Chief of the United States Navy.

List of Illustrations.

Pumping Engine and Pumps—Plan of Dry Dock and Pump-Well – Sections of Dry Dock—Engine House—Iron Floating Gate—Details of Floating Gate—Iron Turning Gate—Plan of Turning Gate—Culvert Gate—Filling Culvert Gates—Engine Bed—Plate, Pumps, and Culvert—Engine House Roof—Floating Sectional Dock—Details of Section, and Plan of Turn-Tables —Plan of Basin and Marine Railways—Plan of Sliding Frame, and Elevation of Pumps—Hydraulic Cylinder—Plan of Gearing for Pumps and End Floats —Perspective View of Dock, Basin, and Railway—Plan of Basin of Portsmouth Dry Dock—Floating Balance Dock—Elevation of Trusses and the Machinery—Perspective View of Balance Dry Dock

---

Free Hand Drawing.

Profusely Illustrated. 18mo. **Boards. 50 cents.**

A GUIDE TO ORNAMENTAL, Figure, and Landscape Drawing. By an Art Student.

CONTENTS.—Materials employed in Drawing, and how to use them—On Lines and how to Draw them—On Shading—Concerning lines and shading, with applications of them to simple elementary subjects—Sketches from Nature.

Minifie's Mechanical Drawing.

Ninth Edition.

Royal 8vo. Cloth. $4.00.

A TEXT-BOOK OF GEOMETRICAL DRAWING for the use of Mechanics and Schools, in which the Definitions and Rules of Geometry are familiarly explained; the Practical Problems are arranged, from the most simple to the more complex, and in their description technicalities are avoided as much as possible. With illustrations for Drawing Plans, Sections, and Elevations of Buildings and Machinery; an Introduction to Isometrical Drawing, and an Essay on Linear Perspective and Shadows. Illustrated with over 200 diagrams engraved on steel. By WM. MINIFIE, Architect. Eighth Edition. With an Appendix on the Theory and Application of Colors.

"It is the best work on Drawing that we have ever seen, and is especially a text-book of Geometrical Drawing for the use of Mechanics and Schools. No young Mechanic, such as a Machinist, Engineer, Cabinet-Maker, Millwright, or Carpenter, should be without it."—*Scientific American.*

"One of the most comprehensive works of the kind ever published, and cannot but possess great value to builders. The style is at once elegant and substantial."—*Pennsylvania Inquirer.*

"Whatever is said is rendered perfectly intelligible by remarkably well-executed diagrams on steel, leaving nothing for mere vague supposition; and the addition of an introduction to isometrical drawing, linear perspective, and the projection of shadows, winding up with a useful index to technical terms." —*Glasgow Mechanics' Journal.*

☞ The British Government has authorized the use of this book in their schools of art at Somerset House, London, and throughout the kingdom.

Minifie's Geometrical Drawing.

New Edition. Enlarged.

12mo. Cloth. $2.00.

GEOMETRICAL DRAWING. Abridged from the octavo edition, for the use of Schools. Illustrated with 48 steel plates. New edition, enlarged.

"It is well adapted as a text-book of drawing to be used in our High Schools and Academies where this useful branch of the fine arts has been hitherto too much neglected."—*Boston Journal.*

Bell on Iron Smelting.

8vo. Cloth. $6.00.

CHEMICAL PHENOMENA OF IRON SMELTING. An experimental and practical examination of the circumstances which determine the capacity of the Blast Furnace, the Temperature of the Air, and the Proper Condition of the Materials to be operated upon. By I. LOWTHIAN BELL.

Battershall's Legal Chemistry.

Illustrated. 12mo. Cloth. In press.

LEGAL CHEMISTRY. A Guide to the detection of Poisons, Falsification of Writings, Adulteration of Alimentary and Pharmaceutical Substances; Analysis of Ashes, and Examination of Hair, Coins, Fire-Arms, and Stains, as applied to Chemical Jurisprudence. For the use of Chemists, Physicians, Lawyers, Pharmacists, and Experts. Translated with additions, including a list of books and memoirs on Toxicology, etc., from the French of A. NAQUET. By J. P. BATTERSHALL, Ph.D., with a Preface by C. F. CHANDLER, Ph.D., M.D., LL.D.

King's Notes on Steam.

Nineteenth Edition.

8vo. Cloth. $2.00.

LESSONS AND PRACTICAL NOTES ON STEAM, the Steam-Engine, Propellers, &c., &c., for Young Engineers, Students, and others. By the late W. R. KING, U. S. N. Revised by Chief-Engineer J. W. KING, U. S. Navy.

"This is one of the best, because eminently plain and practical treatises on the Steam Engine ever published."—*Philadelphia Press.*

This is the thirteenth edition of a valuable work of the late W. H. King, U. S. N. It contains lessons and practical notes on Steam and the Steam Engine, Propellers, etc. It is calculated to be of great use to young marine engineers, students, and others. The text is illustrated and explained by numerous diagrams and representations of machinery.—*Boston Daily Advertiser.*

Text-book at the U. S. Naval Academy, Annapolis.

Burgh's Modern Marine Engineering.

One thick 4to vol. Cloth. $25.00. Half morocco. $30.00.

MODERN MARINE ENGINEERING, applied to Paddle and Screw Propulsion. Consisting of 36 Colored Plates, 259 Practical Wood-cut Illustrations, and 403 pages of Descriptive Matter, the whole being an exposition of the present practice of the following firms: Messrs. J. Penn & Sons; Messrs. Maudslay, Sons & Field; Messrs. James Watt & Co.; Messrs. J. & G. Rennie; Messrs. R. Napier & Sons; Messrs. J. & W. Dudgeon; Messrs. Ravenhill & Hodgson; Messrs. Humphreys & Tenant; Mr. J. T. Spencer, and Messrs. Forrester & Co. By N. P. Burgh, Engineer.

Principal Contents.—General Arrangements of Engines, 11 examples —General Arrangement of Boilers, 14 examples — General Arrangement of Superheaters, 11 examples—Details of Oscillating Paddle Engines, 34 examples—Condensers for Screw Engines, both Injection and Surface, 20 examples—Details of Screw Engines, 20 examples—Cylinders and Details of Screw Engines, 21 examples—Slide Valves and Details, 7 examples—Slide Valve, Link Motion, 7 examples—Expansion Valves and Gear, 10 examples—Details in General, 30 examples—Screw Propeller and Fittings, 13 examples - Engine and Boiler Fittings, 28 examples - In relation to the Principles of the Marine Engine and Boiler, 33 examples.

Notices of the Press.

"Every conceivable detail of the Marine Engine, under all its various forms, is profusely, and we must add, admirably illustrated by a multitude of engravings, selected from the best and most modern practice of the first Marine Engineers of the day. The chapter on Condensers is peculiarly valuable. In one word, there is no other work in existence which will bear a moment's comparison with it as an exponent of the skill, talent and practical experience to which is due the splendid reputation enjoyed by many British Marine Engineers."—*Engineer.*

"This very comprehensive work, which was issued in Monthly parts, has just been completed. It contains large and full drawings and copious descriptions of most of the best examples of Modern Marine Engines, and it is a complete theoretical and practical treatise on the subject of Marine Engineering."—*American Artisan.*

This is the only edition of the above work with the beautifully *colored* plates, and it is out of print in England.

Bourne's Treatise on the Steam Engine.

Ninth Edition.

Illustrated. 4to. Cloth. $15.00.

TREATISE ON THE STEAM ENGINE in its various applications to Mines, Mills, Steam Navigation, Railways, and Agriculture, with the theoretical investigations respecting the Motive Power of Heat and the proper Proportions of Steam Engines. Elaborate Tables of the right dimensions of every part, and Practical Instructions for the Manufacture and Management of every species of Engine in actual use. By JOHN BOURNE, being the ninth edition of "A Treatise on the Steam Engine," by the "Artisan Club." Illustrated by thirty-eight plates and five hundred and forty-six wood-cuts.

As Mr. Bourne's work has the great merit of avoiding unsound and immature views, it may safely be consulted by all who are really desirous of acquiring trustworthy information on the subject of which it treats. During the twenty-two years which have elapsed from the issue of the first edition, the improvements introduced in the construction of the steam engine have been both numerous and important, and of these Mr. Bourne has taken care to point out the more prominent, and to furnish the reader with such information as shall enable him readily to judge of their relative value. This edition has been thoroughly modernized, and made to accord with the opinions and practice of the more successful engineers of the present day. All that the book professes to give is given with ability and evident care. The scientific principles which are permanent are admirably explained, and reference is made to many of the more valuable of the recently introduced engines. To express an opinion of the value and utility of such a work as *The Artisan Club's Treatise on the Steam Engine,* which has passed through eight editions already, would be superfluous; but it may be safely stated that the work is worthy the attentive study of all either engaged in the manufacture of steam engines or interested in economizing the use of steam.—*Mining Journal.*

Isherwood's Engineering Precedents.

Two Vols. in One. 8vo. Cloth. $2.50.

ENGINEERING PRECEDENTS FOR STEAM MACHINERY. Arranged in the most practical and useful manner for Engineers. By B. F. ISHERWOOD, Civil Engineer, U. S. Navy. With illustrations.

Ward's Steam for the Million.

New and Revised Edition.

8vo. Cloth. $1.00.

STEAM FOR THE MILLION. A Popular Treatise on Steam and its Application to the Useful Arts, especially to Navigation. By J. H. WARD, Commander U. S. Navy. New and revised edition.

A most excellent work for the young engineer and general reader. Many facts relating to the management of the boiler and engine are set forth with a simplicity of language and perfection of detail that bring the subject home to the reader.—*American Engineer.*

Walker's Screw Propulsion.

8vo. Cloth. 75 cents.

NOTES ON SCREW PROPULSION, its Rise and History. By Capt. W. H. WALKER, U. S. Navy.

Commander Walker's book contains an immense amount of concise practical data, and every item of information recorded fully proves that the various points bearing upon it have been well considered previously to expressing an opinion.—*London Mining Journal.*

Page's Earth's Crust.

18mo. Cloth. 75 cents.

THE EARTH'S CRUST: a Handy Outline of Geology. By DAVID PAGE.

"Such a work as this was much wanted—a work giving in clear and intelligible outline the leading facts of the science, without amplification or irksome details. It is admirable in arrangement, and clear and easy, and, at the same time, forcible in style. It will lead, we hope, to the introduction of Geology into many schools that have neither time nor room for the study of large treatises."—*The Museum.*

Rogers' Geology of Pennsylvania.

3 Vols. 4to, with Portfolio of Maps. Cloth. $30.00.

THE GEOLOGY OF PENNSYLVANIA. A Government Survey. With a general view of the Geology of the United States, Essays on the Coal Formation and its Fossils, and a description of the Coal Fields of North America and Great Britain. By HENRY DARWIN ROGERS, Late State Geologist of Pennsylvania. Splendidly illustrated with Plates and Engravings in the Text.

It certainly should be in every public library throughout the country, and likewise in the possession of all students of Geology. After the final sale of these copies, the work will, of course, become more valuable.

The work for the last five years has been entirely out of the market, but a few copies that remained in the hands of Prof. Rogers, in Scotland, at the time of his death, are now offered to the public, at a price which is even below what it was originally sold for when first published.

Elliot's European Light-Houses.

51 Engravings and 21 Wood-cuts. 8vo. Cloth. $5.00.

EUROPEAN LIGHT-HOUSE SYSTEMS. Being a Report of a Tour of Inspection made in 1873. By Major GEORGE H. ELLIOT, Corps of Engineers, U.S.A., member and Engineer Secretary of the Light-house Board.

Sweet's Report on Coal.

8vo. Cloth. $3.00.

SPECIAL REPORT ON COAL; showing its Distribution, Classification, and Cost delivered over different routes to various points in the State of New York, and the principal cities on the Atlantic Coast. By S. H. SWEET. With maps.

Colburn's Gas Works of London.

12mo. Boards. 60 cents.

GAS WORKS OF LONDON. By ZERAH COLBURN.

The Useful Metals and their Alloys; Scoffren, Truran, and others.

Fifth Edition.

8vo. Half calf. $3.75.

THE USEFUL METALS AND THEIR ALLOYS, including MINING VENTILATION, MINING JURISPRUDENCE AND METALLURGIC CHEMISTRY employed in the conversion of IRON, COPPER, TIN, ZINC, ANTIMONY, AND LEAD ORES, with their applications to THE INDUSTRIAL ARTS. By JOHN SCOFFREN, WILLIAM TRURAN, WILLIAM CLAY, ROBERT OXLAND, WILLIAM FAIRBAIRN, W. C. AITKIN, and WILLIAM VOSE PICKETT.

Collins' Useful Alloys.

18mo. Flexible. 75 cents.

THE PRIVATE BOOK OF USEFUL ALLOYS and Memoranda for Goldsmiths, Jewellers, etc. By JAMES E. COLLINS

This little book is compiled from notes made by the Author from the papers of one of the largest and most eminent Manufacturing Goldsmiths and Jewellers in this country, and as the firm is now no longer in existence, and the Author is at present engaged in some other undertaking, he now offers to the public the benefit of his experience, and in so doing he begs to state that all the alloys, etc., given in these pages may be confidently relied on as being thoroughly practicable.

The Memoranda and Receipts throughout this book are also compiled from practice, and will no doubt be found useful to the practical jeweller. —*Shirley, July,* 1871.

Joynson's Metals Used in Construction.

12mo. Cloth. 75 cents.

THE METALS USED IN CONSTRUCTION: Iron, Steel, Bessemer Metal, etc., etc. By FRANCIS HERBERT JOYNSON. Illustrated.

"In the interests of practical science, we are bound to notice this work; and to those who wish further information, we should say, buy it; and the outlay, we honestly believe, will be considered well spent." — *Scientific Review.*

www.ingramcontent.com/pod-product-compliance
Lightning Source LLC
LaVergne TN
LVHW020227110826
845151LV00003B/848

* 9 7 8 1 4 2 5 5 3 2 1 1 6 *